Isolation Technology

A Practical Guide

Second Edition

T0174097

Isolation Technology
A Practical Guide
Second Edition

Tim Coles

CRC Press
Taylor & Francis Group
Boca Raton London New York

CRC Press is an imprint of the
Taylor & Francis Group, an **informa** business

CRC Press
Taylor & Francis Group
6000 Broken Sound Parkway NW, Suite 300
Boca Raton, FL 33487-2742

First issued in paperback 2019

© 2004 by Taylor & Francis Group, LLC
CRC Press is an imprint of Taylor & Francis Group, an Informa business

No claim to original U.S. Government works

ISBN-13: 978-0-8493-1944-0 (hbk)
ISBN-13: 978-0-367-39425-7 (pbk)
Library of Congress Card Number XX-XXXXX

Library of Congress Cataloging-in-Publication Data

'Catalog record is available from the Library of Congress.

**Visit the Taylor & Francis Web site at
http://www.taylorandfrancis.com**

**and the CRC Press Web site at
http://www.crcpress.com**

Preface to Second Edition

It is now some five years since the first edition of this book appeared. There have been some technological advances in this time, but perhaps the most significant changes have been in the acceptance and the understanding of isolation technology. This acceptance may best be demonstrated by the series of monographs, guidelines, and standards produced by various bodies, designed to describe best practice in the design and operation of isolators. These include:

- Isolators Used for Aseptic Processing and Sterility Testing, Pharmaceutical Inspection Cooperation Scheme (Europe)
- Design and Validation of Isolator Systems for the Manufacturing and Testing of Healthcare Products, Parenteral Drug Association (USA)
- Sterile Drug Products Produced by Aseptic Processing, Food and Drug Administration (USA)
- ISO EN 14644 Part 7, Separative Devices
- Isolators for Pharmaceutical Applications, 3rd Edition, UK Pharmaceutical Isolator Group
- Handling Cytotoxic Drugs in Isolators in NHS Pharmacies, HSE and MHRA (UK)
- Recommendations for the Production, Control and Use of Biological Indicators for Sporicidal Gassing of Surfaces within Separative Enclosures, PDA Committee (USA and UK)

This second edition includes descriptions of and comments on these new documents. Recent technology — such as the new breed of sanitising gas generators — has been brought into the appropriate chapters, and the text has been updated throughout to reflect more recent thinking. Finally, minor errors in the previous edition have been corrected.

The second edition draws heavily on the content of the guideline booklet, *Isolator for Pharmaceutical Applications — 3rd Edition*, for both information and inspiration. I was closely involved in the assembly and editing of this recent work and make no apology for trawling its content. I am, however, grateful to my co-editors and to the UK Pharmaceutical Isolator Group for allowing me this privilege.

Thanks are also due to GRC Consultants; Malcolm Hughes, Dabur Oncology; Ray Collyer, Dabur Oncology; James Drinkwater, BioQuell Ltd.; and Brian Midcalf, UK Pharmaceutical Isolator Group.

Tim Coles
Cambridge

The Author

Tim Coles holds B.Sc. and M.Phil. degrees in environmental sciences and has been active in the field of pharmaceutical isolation technology for the last 20 years. He has worked with La Calhène SA, Cambridge Isolation Technology, and MDH (now BioQuell Ltd.), and is currently employed as an isolator specialist by GRC Consultants, part of the Mott-MacDonald group. Coles has been active in the UK Pharmaceutical Isolator Group and is on the editorial committee responsible for producing comprehensive new guidelines on isolation technology. He is also a member the PDA Biological Indicator Group and is a frequent speaker at conferences and seminars in the UK and Europe.

Contents

chapter one

Introduction

The background to isolation technology

The concept of the glovebox, used to protect a process from the operator, or to protect the operator from a process, is hardly new. Gloveboxes were first developed with the atomic weapons programme during World War II; development continued within the nuclear power industry, up to the present day. Gloveboxes were also used almost from the beginning of sterile product manufacture, because operators were quickly recognised to be the major source of contamination. The use of gloveboxes declined when reliable panel high efficiency particulate air (HEPA) filters (see Chapter 2) became available (Agalloco 1995). These filters led to the development of cleanrooms, which have dominated sterile production until recently. Gloveboxes have also been developed for nonnuclear containment purposes, particularly where pathogenic organisms are involved, and clear standards exist for such containments. The Class III Biological Safety Cabinet defined in BS 5295, and latterly in EN 12469, is, of course, a glovebox.

It is not easy to establish dates, but around 20 years ago, some subtle changes began to take place in the commercial climate and especially in the pharmaceutical industry. Consumers began to demand improved product quality whilst, at the same time, better standards of safety for those handling potentially hazardous materials became the norm. At about this time, the French company La Calhène SA recognised that some of the products that it had developed, specifically for the nuclear industry, might well have application in other areas, especially pharmaceuticals. In particular, the well-engineered *Double Porte de Transfert Etanche* (DPTE), or *Double-Door Transfer Port*, also known as the *Rapid Transfer Port* (RTP™) (see Chapter 3), which moved highly toxic plutonium oxide powder from one glovebox to another, was seen as a device with a future in pharmaceuticals. A further factor was the work of Professor Philip Trexler in the 1950s, who produced sealed enclosures from the clear polyvinyl chloride (PVC) films that were then newly introduced. These, combined with glass-fibre "candle" air filters, provided simple and cheap contamination-free environments for research animals. Indeed, the success of these enclosures was so great that truly specific pathogen-free (SPF) animals could be reared and maintained for long periods; their use continues widely today.

It has been suggested that specific technical developments have led to the recent expansion of isolation. For instance, Carmen Wagner states:

> The DPTE (RTP) is, in the author's opinion, a milestone in the evolution of advanced aseptic processing isolators. (Wagner 1995)

This is not entirely true, because the La Calhène SA port was freely available 25 years ago, yet the real expansion of isolation has only been in the last ten years. The RTP is a very useful device for use in isolation, but, in truth, it has been a blend of various issues — commercial, social, and technical — which has led to a reevaluation of gloveboxes and specialised enclosures, and to the emergence of a new philosophy called *isolation technology*.

The technology has applications in many areas of endeavour, but it is principally with its application in the pharmaceutical industry, as regulated by the Medicines and Health Care Products Regulatory Agency (MHRA), formerly the Medicines Control Agency, in the UK and by the Food and Drug Administration (FDA) in the U.S., that this book is concerned. This scope extends, however, to include hospital pharmacy work and some aspects of biotechnology, research animals, and direct medical uses. Figure 1.1 shows the development pathways of the various types of isolator applications.

Isolation technology — a definition

It seems only fair to explain what is meant by isolation technology at an early stage in this account, although this seems to be an area where agreement is still lacking. The definition of the technology is bound up with the definition of the word *isolator*. In the recent publication *Isolator Technology*, edited by Carmen Wagner and James Akers (1995), the word isolator is given four separate definitions in the glossary section, reflecting the views of several authors. Isolators are then further subdivided into open and closed categories. Meanwhile, in the guideline booklet *Isolators for Pharmaceutical Applications* (Lee and Midcalf 1994), an isolator is defined as follows:

> A containment device which utilises barrier technology for the enclosure of a controlled workspace.

Isolators are then subdivided in the booklet into Type 1, positive pressure, used for product protection, and Type 2, negative pressure, used for operator protection. This definition was criticised by James Lyda of the Parenteral Drug Association (PDA) (Lyda 1995) for failing to utilise the words *environment* and *microbiological quality*. However, a fully comprehensive definition

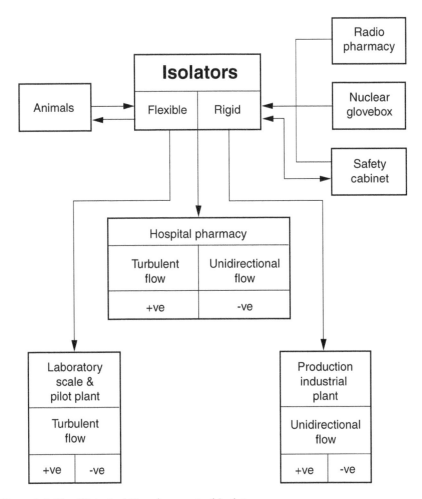

Figure 1.1 The Historical Development of Isolators.

would become inordinately lengthy, and these issues are clearly addressed elsewhere in the booklet.

Yet another definition has been given by the working group established by the International Organisation for Standardisation (ISO) on contamination control (ISO/TC209/WG7, reported by Brammah 1995), as follows:

> A localised environment created by a sealed enclosure to isolate the product from contamination and/or people.

This committee has more recently produced Part 7 of ISO 14644 (ISO/FDIS 14644–7 2001), in which an isolator is defined as "An industry specific separative enclosure." In light of such arguments, it seems that, for the purposes of this book, a rigorous definition is probably counterproductive.

Let us say, then, that isolation technology is the placement of a physical barrier between a process and its operators. The purpose of the barrier may be to protect the process and its materials from the effects of the operators, or it may be to protect the operators from the effects of the process; in some cases, it may seek to do both. The barrier may be total, so that the process is always behind either a physical wall or at least behind HEPA filters, or it may be partial, so that the process may be separated only, for instance, by engineered airflow.

Beyond this, we can say that the environment, workspace, or critical area, delineated by the barrier, should have a defined quality that takes account of the intended purpose. This quality may be defined in terms of microbiology, particle burden, humidity, oxygen content, or whatever combination is appropriate.

Isolation technology versus barrier technology

The use of the word *barrier* in the preceding section promptly engenders a new argument concerning the differences, be they technical or semantic, between isolation technology and barrier technology. It is probably true to say that the terms were practically synonymous in the early days, but now it seems to be accepted that they do convey quite separate technical meanings.

James Agalloco clearly described the received wisdom in a paper given to a meeting of the Scottish Society for Contamination Control in Birmingham, UK, in 1996 (Agalloco 1996). Put quite simply, isolation technology is absolute and barrier technology is not. In true isolators, a physical wall of perhaps plastic or stainless steel exists at all times between the process and the operators. The inlet and exhaust of air, or other gases, can be only via HEPA filters. This includes any transfer devices in use on the isolator.

By contrast, barrier technology allows for some limited exchange of atmospheres between the workspace and the outside environment. Thus, a rather open form of barrier is the curtaining around a Vertical Laminar Flow unit in a cleanroom, whilst a more closed form would be the mousehole devices frequently used for the exit of vials from enclosed filling lines (see Chapter 3). Carmen Wagner (1995) describes partial barriers and closed barriers, along with open isolators and closed isolators, under the all-embracing title of *Protective Barrier Systems*. The closed isolator has full containment and is probably used only in batch processes, whereas the open isolator may carry a transfer device, like the mousehole, with very restricted atmospheric exchange, inevitable for continuous line processes. The degree to which the defined workspace is closed has implications for the establishment and maintenance of a sterile environment. Clearly, a sealed isolator is more easily sterilised (though note the limitations of the word *sterile*, which are discussed at the start of Chapter 7) than a barrier enclosure that may exchange air with the outside environment. As a rule, this book is concerned only with isolation technology; but under the strict definition, certain aspects of barrier

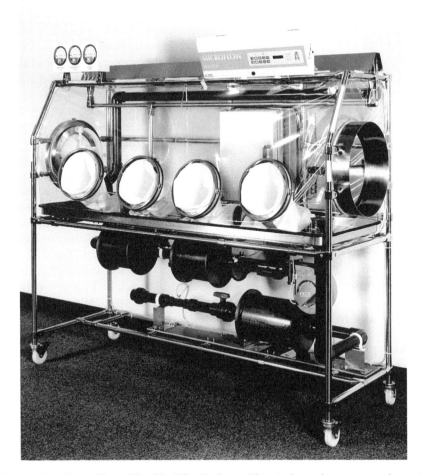

Figure 1.2 A Four-Glove Flexible Film Isolator. This isolator forms part of a suite used for aseptic dispensing in a large hospital pharmacy and is used as a sterile bank, storing materials ready for dispensing in other smaller isolators. Note the RTP on the left-hand end and the refrigerator fitted to the back wall of the isolator. (Courtesy of Astec Microflow.)

technology may be invoked, particularly in continuous process applications. Figure 1.2 shows a typical flexible film isolator with many of the features described later in this book.

The aim of isolation technology

There are a number of factors that may combine in various ways to lead the user toward isolation technology. The priority of such factors will depend on the nature of the process, but, generally speaking, the aims of the technology are given below.

Improvement in product quality

Sterile products, such as parenteral preparations (para = beside, enteron = gastrointestinal tract — essentially this means injectable drugs), form a large part of the production of pharmaceutical companies. Some of these preparations may be sterilised at the end of production (terminal sterilisation) by autoclaving or gamma irradiation, but many products are too sensitive to withstand such treatments. Instead, they can be produced only by sterile or aseptic manufacture, as carried out in cleanrooms. However, even in the most technically advanced and well-run cleanrooms, there is always a potential for the large number of particles shed by operators to reach, and thus contaminate, the product. Unlike particles produced by equipment, operator-shed particles are quite likely to carry viable organisms and so are very undesirable. By placing the process inside an isolator, the potential for operator-spread contamination is virtually eliminated, and, thus, the sterility of the product is enhanced. It is worth considering the concept of Sterility Assurance Level (SAL) at this point. The SAL of a product is defined as the probability that it may contain viable organisms. Thus, an item may be said to have an SAL of 10^{-4}, in which case one might expect to test 10,000 units before finding one containing viable organisms. Another title recently suggested for the same concept is *probability of a nonsterile unit*, or PNSU.

As reported by James Akers (Akers 1995), a great deal of dogma has developed around the concept of sterility in the pharmaceutical industry, with little technical data to support any particular view. A general overview has been that aseptic processing results in products with an SAL of 10^{-3}, whilst terminal sterilisation results in an SAL of 10^{-6}. Akers suggests that even the best isolator systems cannot reach the same SAL as terminal sterilisation, but that isolators may reach an SAL of 10^{-6}, whilst terminal sterilisation in fact well exceeds the 10^{-6} level. This issue needs to be resolved by validation for any individual sterile production project, but the evidence is that the use of isolators can result in a much higher SAL than the conventional cleanroom. This view, however, is not wholly embraced by all parties and, in particular, prospective isolation users should consider the logic offered by the MHRA in Chapter 10.

This increase in SAL stems from two factors: (1) the internal environment of the isolator can potentially be of a very high standard and (2) the isolator may be sterilised to a very high standard prior to use. These conditions may be maintained during the operation of the process, whereas in the cleanroom, standards inevitably fall with the entry of operators. The very important issue of isolator sterilisation is covered in Chapter 7.

Whilst sterility is probably the major product quality factor in pharmaceutical isolation technology, other quality issues may also be addressed. For instance, an isolator may be run with an atmosphere that benefits the product, such as low relative humidity or inert gas, both of which would be difficult to achieve in cleanroom conditions. In addition, there may be overall quality benefits in terms of cross-contamination or exposure to general cleaning materials and the like.

Reduction in costs

The capital cost as well as the running cost of the conventional cleanroom designed for aseptic production is high. Very often, a dedicated building structure will be required, fitted out with highly specialised internal constructions and containing heating and ventilation systems capable of handling very large quantities of air. For example, a cleanroom to house a large aseptic filling line might measure 50 m long, 10 m wide, and 2.5 m high. This has a volume of 1,250 m³. If vertical laminar (unidirectional) downflow is provided over the whole area at a velocity of 0.40 m/sec, then the room volume will be changed every 6.25 sec, resulting in a flow rate of 720,000 m³/h, weighing a surprising 850 tonnes. A high proportion of this flow will be recirculated, but even so, this is a lot of air to condition and to pass through HEPA filters. Clearly, if the critical volume is reduced to just the few cubic metres of an isolator surrounding the filling line, then there is considerable potential to reduce plant costs.

Similar arguments apply to running costs. The cleanroom will require much energy to condition and move the large flow of air, while the necessary gowning is an expensive operation both in terms of the materials used and the time involved. The continuous work required to clean and monitor cleanrooms also consumes funds, as does the general maintenance of the plant.

Having put forward this point of view, a brief survey reveals some figures that may be surprising with regard to capital costs. Didier Meyer (Meyer 1995) states that, in some production applications, capital expenses for isolation may, in fact, be more than for a conventional cleanroom. Henry Rahe (Rahe 1996) suggests that, while savings of up to 40 percent may be realised in some cases, isolators may indeed be more expensive than cleanrooms. Similar information is given by Gordon Farquharson, who states:

> In the case of aseptic filling of pharmaceuticals, it is common for the total facility and process equipment capital expenditure for an isolator application to be 70 to 90 percent of that for a traditional cleanroom system. (Farquharson 1995)

The apparently high capital costs of isolators here is probably the result of at least four factors:

1. Special, on-off design and construction: Whilst there are plans for the duplication of some successful, isolated production lines, each pharmaceutical production isolator is a completely new project. This means that design work, any developments required, tooling and

jigs, construction, assembly, installation, commissioning, and validation all have to be recovered in the costs of the one system. Design
work, in particular, can be a lengthy and, thus, expensive process.
Clearly, these factors combine to produce what appears at first glance
to be a high price.

2. Integration with other equipment: The fact of isolation often means
    very intimate connections of the isolator with process equipment
    of various types. These could, for instance, be depyrogenating tun-
    nels, autoclaves, freeze dryers, and, of course, filling machines.
    Careful liaison will be needed with the equipment manufacturer to
    achieve a successful union. Since the equipment manufacturer may
    well be in a different country, the costs of the integration process
    can be high.

3. Fabrication and finishing costs: Isolation systems in production ap-
    plications, either sterile or toxic containments, are generally con-
    structed from stainless steel sheets (see Chapter 2). These fabrications
    are often quite complex in geometry, in order to house the operational
    equipment and to allow for perhaps unidirectional downflow and
    clean-in-place (CIP) features. Furthermore, these structures must be
    built to sanitary standards, with generously radiused edges and
    ball-formed corners throughout. All of the forming and welding must
    be dressed and polished, very often not only on the inside but also
    on the exterior — a labour-intensive process. Thus, the costs of the
    basic structure can be apparently high.

4. Instrumentation and control: Because the environmental emphasis is
    now focused on the isolator and not the room, most of the control,
    instrumentation, and monitoring will be applied to the isolator and
    hence appear as part of its cost. This can include extensive instru-
    mentation of pressure and flow throughout the system, with control
    of a number of fans and even a certain amount of integration with
    process control. All of this may be under the command of a program-
    mable logic controller (PLC), for which software must not only be
    developed but also be fully validated if the isolator is to be used for
    licensed manufacture, all of which is a fairly expensive exercise.

The running costs, however, for isolator systems are much lower than
for the conventional cleanrooms. Direct power consumption is normally
confined to one or two fans, which may be a few hundred watts each in the
case of turbulent isolators and a few kilowatts in the case of laminar flow
isolators, which move much more air. The indirect power consumption of
air-conditioning systems is much less, since only relatively small volumes
of air are now involved: hundreds of cubic metres of air per hour instead of
hundreds of thousands. The time and expense of gowning the operators is
removed and is primarily replaced by the cost of changing gloves. Cleaning
and sanitisation costs are significantly reduced, both in terms of materials
and labour, while the costs of environmental monitoring, such as by settle

plates, are also greatly reduced in isolators. Lastly, the general maintenance of isolators, planned and incidental, is cheap and quick when compared with the extensive engineering of cleanrooms.

Improvements in safety

Sterile production is a major use of isolation, but the containment of toxic processes is at least as large and potentially much larger. More active compounds are being developed through the very extensive research of pharmaceutical and biotechnology industries, and these have to be transported, charged into reactors, filtered, dried, blended, and packaged. This may take place on a large industrial scale or on a small laboratory scale. Air suits and powder booths will serve well in many applications, but there is increasing pressure to reduce the hazards to operators and to the environment. Isolation of the process can reduce the levels of contamination to extremely low values if required; indeed, it is normal to quote containment in terms of nanograms of active substance per cubic metre of room air.

The production of the new generation of anticancer cytotoxic drugs provides a very good example. Here, there is a simultaneous need for the containment of some very active compounds and for the maintenance of an aseptic environment for processing and filling. Isolation technology provides a solution to both problems, although the conundrum of positive- or negative-pressure containment for such application is yet to be resolved (see Chapter 5). Despite the operational pressure issue, the use of isolators has become more or less standard in the production of cytotoxic parenterals.

Flexibility

The building used to house a cleanroom is complex and specialised and, therefore, takes a long time to construct and commission. The housing requirements of an isolator system, however, even for a complex production line, can be very much less demanding (see Chapter 5). Thus, an existing building may potentially be fitted out with isolators and commissioned quickly, without major forward planning and construction schedules, though experience suggests that the lack of knowledge of isolation limits this effect.

Furthermore, if the demands of the site change, then the isolators may be removed and reinstated elsewhere, for comparatively little cost. By contrast, cleanrooms are effectively permanent structures that, once completed, can be used for little other than the intended purpose. Thus, isolation technology relieves some of the heavy burden of resource commitment, allowing more efficient use of space and time.

Special conditions

A significant number of pharmaceutical, biotechnological, and other industrial processes have special environmental requirements other than freedom from particulate contamination. It is common to find that an inert atmosphere, such as a nitrogen, low-humidity, or low-oxygen atmosphere, is needed to prevent deterioration of a material or product during processing. Clearly, containment of the work inside a suitable sealed enclosure with appropriate atmospheric control is the way forward: isolation technology at work once again.

Qualification

Since an isolator system is likely to be physically much smaller than the equivalent cleanroom arrangement, it can be argued that qualification (or validation) — installation qualification (I Q) and operational qualification (OQ) — should be simpler and quicker. Clearly, there will be fewer filters to test and fewer sample points for, say, particle counting and airflow testing. In practice, however, isolators seem to take as long, if not longer, to qualify than classical cleanrooms. This puzzling issue is further discussed in Chapter 8.

Aesthetics

The visual appearance of any piece of equipment, from a car to a coffeepot, will affect the way in which the user handles it and the way in which an observer views it. In the early 21st century, we expect to see not only outward signs of good functionality in our industrial designs but also conformity to modern concepts.

Isolation technology carries with it the connotations of very advanced, space-age design, and we can capitalise on this notion by paying attention to aesthetics. Then the operators will feel that they have the best available equipment to perform their work, whilst the regulatory authorities will feel, albeit perhaps subliminally, that the equipment is right for the job even before they start to look at the process in detail.

Some typical applications of isolation technology

The areas where isolation technology has the most to offer are low volume, high environmental requirement, and high value, but it has very effective applications in areas where these parameters do not specifically apply. Indeed, many industrial processes might benefit from the technology, but it is in the pharmaceutical industry in particular that isolation technology is beginning to make real advances. The following is a brief outline of some typical examples.

Sterility testing

All pharmaceutical products sold as sterile must by law be tested for sterility. This means that a statistically significant number of samples must be removed from each production batch and tested for the presence of micro-organisms. This may be done by direct inoculation of the product into a suitable growth medium, or by filtration of the product and placement of the filter into a growth medium. In either case, incubation is carried out for 14 days and, if no growth is evident, the batch is cleared for distribution and sale. If growth appears, then further tests must be carried out to establish if the positive growth is an artefact of the test procedure, a false positive, or if growth is due to a real contamination problem somewhere in the processing. During this time of retest, the entire batch of product is held in quarantine, and this may represent a large capital sum. Clearly, the false positive is the bane of the quality assurance (QA) microbiologist's life. This situation demands the best possible conditions for testing; otherwise, the process of testing may simply contaminate samples that were actually sterile. The use of isolators for sterility testing has proven to be a classic application of the technology, with users reporting a virtually zero false-positive rate, to the extent that some companies immediately reject the product batch if any positive growth shows up.

A common arrangement for this type of work is a half-suit, flexible film, positive-pressure isolator coupled to a two-glove isolator. The larger isolator is used to carry out the work, and is maintained sterile, whilst the smaller isolator is used as a lockchamber to transfer materials and samples in and out of the system. Gas phase sterilisation (see Chapter 7) is used to sanitise each batch of work in the glove isolator. The actual testing may be by direct inoculation, conventional membrane filtration with a multipoint manifold, or by proprietary methods, such as Millipore Corporation's Steritest. A recent and very sophisticated refinement of the latter process is the version designed specifically to be built into the base tray of isolators, called Steritest Integral™. This places the pump drives and control outside the isolator, with just the stainless steel pump head accessible inside the isolator. Such a system may be quickly installed in a QA microbiology laboratory with little change to a standard room, and it can produce excellent results following validation. Figure 1.3 shows a sterility-testing isolator system based on flexible film technology.

Aseptic filling

Very large numbers of bottles, vials, and ampoules are filled at high speed by automatic machinery in the production of pharmaceuticals. Where they are required to be sterile and cannot be terminally sterilised, they must be filled in completely sterile environments, a process known as aseptic filling. Traditionally, cleanrooms have provided these conditions, but as discussed earlier in this chapter, the industry now has stricter standards. Thus, isolators are being fitted to aseptic filling lines to provide a localised, high-grade environment.

At the smaller end of the scale, isolators are used to manually or semi-manually fill batches of product for research or clinical trials. A few

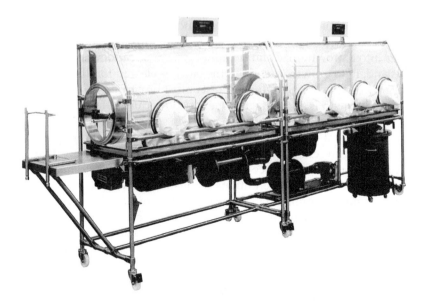

Figure 1.3 A Flexible Film Isolator System Used for Sterility Testing in a Blood Product Laboratory. The left-hand isolator is used to sterilise and transfer materials through to the working isolator on the right. Note the "railway" system used to move sample trays easily through the isolators, and the RTP wastebin underneath the working isolator. (Courtesy of Astec Microflow.)

hundred units are produced at a time, in single batches rather than as continuous runs. The isolator is useful here because the work may then take place under excellent conditions within the research laboratory area, rather than occupying valuable production space in a cleanroom that may be some distance away.

Relatively simple and inexpensive, standard, flexible film isolators may be used, together with a small gas generator for sanitisation of the system, to carry out this work. Once validated, the equipment can be used for a variety of products and processes, provided that precautions are taken to avoid cross-contamination of products.

At the industrial end of the scale, isolators are employed to fill containers at rates of tens of thousands per hour. These isolators are complex and usually specialised, interfacing not only with the filling machine itself, but also with the means to input the product, the containers and their closures, and the means to output the filled product and any waste. They are generally designed and built in close conjunction with the suppliers of process equipment, which may well be adapted specifically for isolated operation. Such isolator systems are usually fabricated from stainless steel sheets and fitted with glass or polycarbonate (Makrolon™ or Lexan™) windows. The filling machines may handle bottles, vials, bags, or ampoules, and, in the latter case, the ventilation equipment must take into account the heat and water vapour generated by the oxy-gas sealing flames.

Isolated filling lines constitute major projects that should embrace not only the isolator manufacturer, but also process equipment manufacturers, the QA department, building engineers, the regulatory authority, and even process operators, at the very earliest stages of design (see Chapter 5). Figure 1.4 illustrates a stainless steel isolation system built onto a filling machine.

Toxic containment

Since the very nature of the business is biologically active materials of one sort or another, it should come as no surprise that there are many areas of pharmaceutical production that require protection from toxic hazards. A typical example is that of break-of-bulk. Active materials are often supplied in bulk form, perhaps in kegs of 25 or 50 kg or even larger intermediate bulk containers (IBCs), and this material must be taken from the bulk

Figure 1.4 A Special Stainless Steel Isolator Fitted onto a Vial Filling Machine. The large stainless steel tubes are the return ducts for the recirculating unidirectional airflow system. Note the RTPs fitted to the end and back walls for the transfer of materials and products. (Courtesy of La Calhène SA.)

container and weighed out, more or less accurately, into smaller batches for addition to further processes. This work is usually manual and often messy as well as hazardous, but a suitably designed isolator can solve a lot of the problems.

Such an isolator might be based on a half-suit, in which the operator carries out the work. A lockchamber is provided for the input of the bulk product, using mechanical handling, such as a pallet truck, if required. Empty, sub-batch containers also go through the lockchamber, and weigh scales are incorporated for the bulk container and the batch containers, perhaps with connection through to the computer handling overall production. The operator moves materials from the bulk to the smaller containers as required, and these may leave via a lockchamber, an RTP, or a product pass-out port (see Chapter 3) at any time. When the entire process is complete, the isolator and its contents are decontaminated, probably by washing down with a deactivating solution piped directly into the isolator, and the empty bulk container is removed, to be replaced with the next one. If the isolator is mobile to some degree, then it can be used for a variety of processes on the site, increasing its cost-effectiveness accordingly. Figure 1.5 shows a stainless steel process containment isolator.

Hospital pharmacy work

The pharmacy department in a large, modern hospital dispenses a wide variety of drugs in many different formats, a good proportion of which are in the form of infusion bags. These divide broadly into antibiotic preparations, total parenteral nutrition (TPN — intravenous nourishment for patients unable to take food by mouth), and cytotoxics (anticancer chemotherapy compounds). All of these must, of course, be prepared under sterile, aseptic conditions and, in the case of cytotoxics, the operator preparing the infusion must also be protected. Various laminar flow devices and small cleanrooms have been used for this work, but in the last few years there has been a move toward the use of isolators in hospital pharmacies to improve both patient standards and operator safety. Figure 1.6 illustrates a negative-pressure containment isolator designed for hospital pharmacy application.

The form in which the technology is applied in hospital pharmacies is still subject to considerable debate, but there is a broad consensus in the UK that cytotoxic dispensing should be handled in rigid, negative-pressure, laminar downflow isolators. TPN and antibiotics may be handled in flexible film, positive-pressure, semi-laminar or turbulent flow isolators. Generally speaking, sanitisation is achieved by simple hand spraying of agents such as alcohol, rather than by more sophisticated gassing systems. None of these arrangements are hard-and-fast rules as yet, and, although generally good results are being obtained, a form of standardisation will hopefully be developed over the next few years.

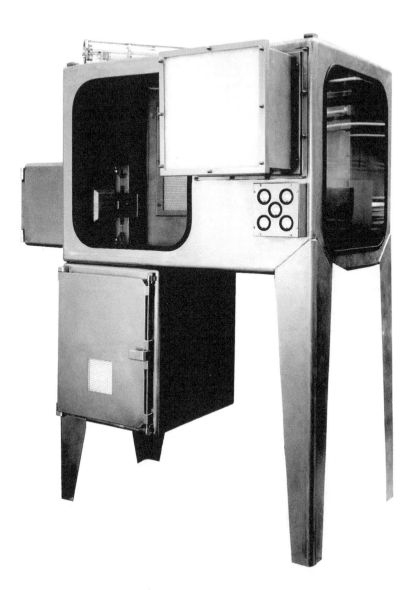

Figure 1.5 A Special Stainless Steel Half-Suit Isolator Built for Toxic Process Containment. (Courtesy of Astec Microflow.)

Another aspect of hospital pharmacy work is the preparation of radio-pharmaceuticals for blood labelling and other diagnostics. More specialised isolators have been developed for this work, which include appropriate radiological shielding and built-in required equipment, such as technetium columns and centrifuges. Figure 1.7 illustrates an isolator designed for nuclear medicine application.

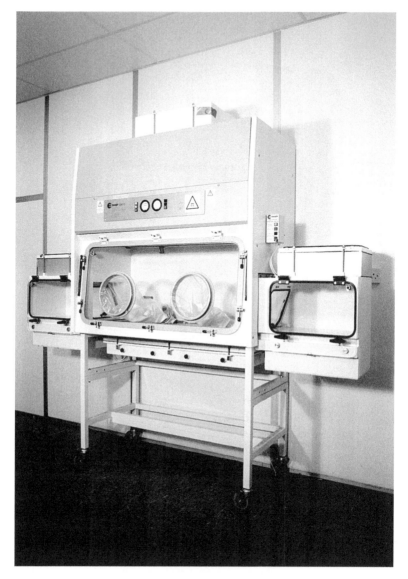

Figure 1.6 A Rigid Wall Isolator. This isolator has unidirectional downflow in negative pressure. It is designed for use in a hospital pharmacy. (Courtesy of Envair Ltd.)

Biomedical isolation

The area of research animal accommodation, particularly of rodents, has long
used isolation technology. Clearly, random influences should be removed as
far as possible from work such as drug trials in order to distinguish the effect
of a test from other factors. For this reason, special strains of research animals
have been developed and reared in rigorous conditions to be SPF. These may

Figure 1.7 A Rigid Wall Isolator. This isolator has unidirectional downflow in negative pressure. It is built for use in nuclear medicine. Some biological shielding is provided in the unit for operator safety. (Courtesy of Envair Ltd.)

then be housed in isolators during the period of the research to eliminate the effects of random infection.

Another useful research animal is the nude mouse, a variety that lacks an immune system. This animal can be used as a host for tissues from other species, but, with no immunity, it must be kept in the infection-free atmosphere of a positive-pressure isolator. Rodents are also used in work on human pathogens, such as human immunodeficiency virus (HIV), tuberculosis (TB), and hepatitis. Because they are then carriers of the disease, they are housed in negative-pressure containment isolators to protect the workers. Transgenic animals are likewise housed in negative pressure. Flexible film isolators of fairly simple design, both positive and negative pressure, have been used extensively for this kind of work in the past 25 years, probably representing the largest single use of isolators to date. Their use seems set to continue and their sophistication to increase as biotechnology and drug development move forward. Figure 1.8 shows a series of biomedical isolators in use.

Surgical and other miscellaneous uses of isolation technology

Surgery is, of course, an activity in which a sterile atmosphere is important and, in the case of deep and extensive surgery such as joint replacement, it is vital if postoperative infection is not to set in with grave risk to the patient. Trials have been carried out in which a sterile, flexible film isolator was lowered onto the operation site and the surgeon worked inside this isolator. Thus, wounds were open only in a sterile atmosphere. The results were very good in terms of postoperative infection rates, but the technique has not been widely adopted. This may be because surgeons have found the isolator to be a considerable encumbrance; but with better design, and with the new risks of physician infection from diseases such as acquired immune deficiency syndrome (AIDS) and hepatitis, surgical isolation may yet become common practice. Other areas in which simple and cheap sterile environments for medical applications could prove effective include third world countries and military battle zones.

Food technology is advancing steadily, and trials have taken place with aseptic food handling. It is possible that future food preparation lines may include some aspects of isolation technology to reduce bioburden as much as possible and, hence, increase shelf life.

Semiconductor production is a technology that demands the very highest levels of contamination control, far beyond those specified for pharmaceutical applications. Whilst the pharmaceutical industry is largely content with a top-grade atmosphere of ISO Class 5, semiconductor fabricators are now thinking in terms of ISO Class 2, which is three orders of magnitude cleaner. Maintaining these conditions in the manned state seems hardly practical; thus, forms of isolation are likely to develop in this field. The nature of the wafer fabrication process does, however, raise major challenges for isolation, since it combines some varied techniques, such as wet chemistry, high temperature furnaces, and high precision optical equipment. Figure 1.9 is an example of flexible film isolators in use for general aseptic processes.

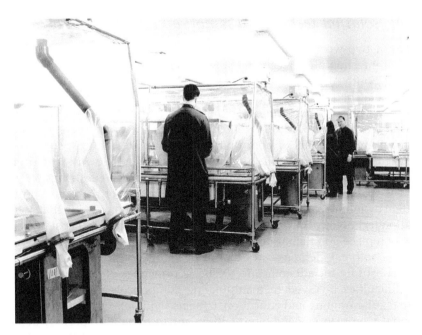

Figure 1.8 A Number of Large Biomedical Isolators in Use at a Research Facility. (Courtesy of Astec Microflow.)

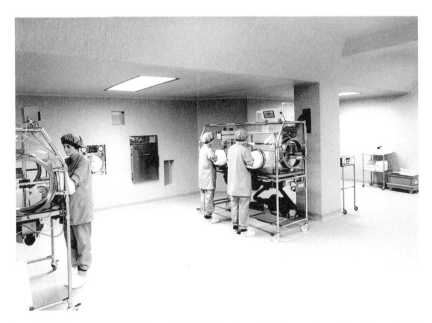

Figure 1.9 Flexible Film Isolators in Use for Aseptic Dispensing. (Courtesy of Nova Laboratories, Ltd.)

What isolation technology is not: a panacea

During a two-week period in September 1992, eight children died from massive infection after receiving TPN at four different hospitals in Johannesburg, South Africa. The source of this infection was traced to four batches of TPN that had been compounded in flexible film isolators by a private company. The subsequent investigation revealed that, while the equipment and the operating procedures were generally fairly sound, there was a catalogue of minor events of poor operating procedure and QA, such as the failure to clean up fluid spillages. These combined to allow the buildup of high levels of pathogenic organisms in some of the dispensed bags of TPN.

It is worth recounting in full the final paragraph of the report on this incident by John Frean, which duly serves as a profound warning that, whilst isolation technology may produce better products at reduced cost, it can in no way replace good laboratory practice.

This outbreak demonstrates that even the use of proven technology is not a substitute for adherence to the basic principles of aseptic technique. The difficulties involved in manipulating components inside the isolator make regular training and education in the use and upkeep of the equipment essential. Complacency and blind faith in any technology, especially in such a sensitive area as TPN production, must always be avoided (Frean 1996).

A similarly tragic event took place in a hospital in Manchester (UK) in 1993. Any potential users of isolators for sterile application would be well advised to read the report on this incident before starting work (Farwell 1994).

chapter two

An introduction to the technology

The materials

Various materials can be used to construct isolators, but in practice the majority are probably made of stainless steel, with a smaller number made with a variety of types of plastic. The commonly used materials are given below.

Flexible film plastic

As mentioned in Chapter 1, flexible film plastic developed largely as a result of the work of Professor Philip Trexler, who made flexible film isolators to house research animals, particularly mice and, to a lesser extent, rats, during the 1950s. The flexible film forms a sealed envelope, perhaps including the floor, often referred to as the canopy. This is then tied to a rigid framework structure with Velcro® or zip ties, and the whole assembly is mounted on some form of table or trolley. The material used is invariably PVC, with a fairly high plasticiser content to give flexibility and elasticity. It is usually 0.30 to 0.50 mm in thickness and one of three types:

1. Natural. This is a translucent material available in roll form. It is cheap and easily worked, often used for sleeves and canopy ties in animal isolators.
2. Clear. This is a transparent material, but with a slight orange-peel appearance, and is also available in roll form. It is fairly cheap and easily worked, and is used for the general structure of isolator canopies.
3. Optically clear. This truly clear material is produced by pressing sheets of transparent film between finely finished plates, hence the term *press-polished*. It is more expensive and only available in sheets of limited size that then require more joins to form a canopy. Optically clear material is usually confined to the working faces of glove isolators.

Flexible film PVC can be joined by bonding and by ultrasonic welding, but the most reliable method of fabrication is by radiofrequency (RF) welding. The two sheets to be joined are made to form the dielectric in what is, in effect, a large capacitor, normally tuned to a frequency of 27 MHz, though other frequencies can be made to work equally well. Application of power to the circuit causes the materials to fuse where they touch, forming a strong and homogenous bond. A pressure of about 50 lb/in.^{2} (psi) of weld is required to maintain intimate contact, and special tools are needed to make line welds, corners, and circular welds. Carefully designed tooling can even produce three-dimensional welds, which can be useful in fabrication. The simple, primary weld produces a step in the fabrication that, if it were required to remain a good seal under stress, say, around an inlet air duct, could possibly leak. Thus, the secondary process of "blocking down" is vital where the mechanical strength of seal is required, and it is desirable for cosmetic reasons on most RF welds. This simply involves placing the weld line under flat plates in the RF welder so that the weld is flattened down to the same thickness as the parent material. A good RF weld line is almost indistinguishable to the touch and visible only as a translucent line about 5 mm wide. The process of RF welding is fairly simple, but the equipment required, the development of tools, and the tailoring of structures usually makes this a specialist occupation.

Flexible film canopies appear somewhat insubstantial, on first sight, for the main structure of an isolator in what may be a critical application, but they actually are really very practical devices. They are surprisingly tough and resilient and will withstand high pressures, in excess of 1000 Pa, both positive and negative, if correctly tied to their framework, although such pressures may damage the shape by excessive, permanent stretching. A well-made canopy will remain intact despite years of heavy use and is really only susceptible to two forms of attack:

1. Sharps. Flexible film has little resistance to needle pricks and to scalpel cuts.
2. Solvents. Being a fairly soft organic material, flexible film PVC may be damaged by exposure to organic solvents. These may leach out the plasticiser and result in cracking, or they may cause clouding and crazing of the surface. However, it is common practice in hospital isolators to use 70 percent ethyl or isopropyl alcohol for sanitisation and, provided this is only carried out perhaps once per day, there seems to be no problem with flexible film. If the user has any doubt about the resistance of flexible film to a specific agent, then tests should be applied on scrap material.

If a flexible film canopy is damaged in some incident, then a quick, temporary repair can be made with adhesive PVC tape sufficient to complete a batch of work. A more permanent repair can be made by using a patch of

PVC and cyanoacrylate adhesive (e.g., Loctite 406™). This may form a strong, leak-tight repair but is unsightly. Thus, the user may wish to return the canopy to the manufacturer for repair by replacement of the damaged panel. In the case of extensive or accumulated damage, or in the case of a massive contamination with toxic material, the canopy can usually be changed quite quickly and at no great expense.

A further consideration for flexible film PVC is its capacity to absorb some organics, particularly sanitising agents, such as hydrogen peroxide vapour. These are then released over time by degassing, which might affect some processes. In practice, although the canopy may absorb significant quantities of material, the ventilation system keeps the actual concentrations down to very low levels when the isolator is in operation.

Flexible film does have some major advantages, as follows:

- Cost. Being relatively cheap and fairly easy to fabricate, flexible film canopies are an inexpensive way to produce a sealed enclosure.
- Clarity. Since the entire structure of the isolator can be made from transparent or even optically clear material, the operator and any assistants or students can see the contents of the isolator from most sides, and no special illumination is required.
- User comfort. Since the walls of the isolator have a certain amount of give, the flexible film glove isolator is found to be more user-friendly than its rigid wall equivalent. In addition, the clear walls make the use of both glove and half-suit isolators less claustrophobic and more open in aspect.
- Ease of installation. Clearly, even a large flexible film canopy can be folded and easily carried into the isolator room through narrow doors and corridors. Rigid isolators can, by contrast, be very difficult to install.
- Leak-tightness ("arimosis" — see Chapter 8, "Physical Validation"). Experience indicates that it is easier to make a flexible film isolator leak-tight than an equivalent rigid isolator.

Early flexible film isolators were built with a complete flexible film enclosure, including the floor, which was not entirely satisfactory, especially for negative-pressure use where the floor would tend to lift. More recent versions have stainless steel or rigid plastic base trays sealed to a bell-type canopy. These have found wide use in laboratory and other small-scale applications, such as sterility testing.

Rigid plastics

Rigid plastics have the advantage of much greater physical strength than flexible films, but are still fairly cheap and, of course, do not need a support framework. Clear material is generally used, of the following types:

PVC

PVC is a cheap material and is easily formed and welded into isolator chambers, lockchambers, and the like. The clear material does, however, have a strong blue tint that may not be acceptable in some applications. In addition, PVC is very soft and easily scratched, though minor damage can be polished out if necessary. A minor disadvantage of PVC is the difficulty in leak testing with Freon™ (see Chapter 6), since the chlorine content of the plastic registers on the halogen detector. Like flexible film, it may also be affected by solvents.

Acrylic

Also known as Perspex™ and Plexiglas™, acrylic is truly clear and can be formed and solvent welded into a variety of structures, though this is best undertaken by specialists. If a number of identical units are required, then it may be cost-effective to produce tooling and mould acrylic sheets to form isolator shells. The material itself is fairly expensive and the costs of good-quality fabrication are quite high, but the results can perform well and look good. As with flexible and rigid PVC, some solvents may affect acrylic and produce clouding or crazing.

Polypropylene

Polypropylene is not clear, but the white form is occasionally used for some isolator structures since it is cheap, easily formed and welded, and chemically resistant.

Polycarbonate

Also known as Lexan™, polycarbonate is a very tough, clear plastic that can be used to fabricate isolators, but not too easily. Its use is generally confined to the windows of rigid, stainless steel isolators.

Stainless steel

Stainless steel is the material of choice in most production and other arduous isolator applications. It can, of course, be folded, formed, welded, and dressed into complex but clean shapes, although this is a specialist activity if high standards are to be maintained. Many subcontractors offer stainless steel fabrications, but few are able to produce the standard of work required by the pharmaceutical industry. One of two grades of material is generally used, with designations from various countries as given in Table 2.1.

The American Iron and Steel Institute (AISI) designations of 304 and 316 are almost universally used. Grade 304 is perfectly adequate in most applications, with the more expensive grade 316 only used where there is direct

Table 2.1 Stainless Steel Designations

United Kingdom	United States	Germany	France	Sweden
BSI 449 970	AISI	Werkstoff No.	AFNOR	SIS
304S15 EN58E	304	4301	Z6CN 18-10	2333
316S16 EN58J	316	4401	Z6CND 18-12	2343

product contact. There is, however, a general trend to using grade 316L for the entire structure of isolators. The *L* denotes low carbon content, producing an alloy that is less prone to corrosion along weld lines. Note that proprietary test kits ("molybdenum tests") are available for checking whether a given piece of stainless steel is grade 304 or 316.

A sheet thickness of 1.5 mm is most often the minimum used for small structures such as lockchambers, while up to 5 mm may be used for larger structures. As a broad generalisation, the edges of fabrications should have a minimum radius of 25 mm, and corners should be ball formed at the same radius. All welds must be crevice-free and fully dressed before the finishing process begins. Modern techniques, such as pulsed Tungsten Inert Gas (TIG) welding, have made high-quality complex fabrications relatively easy in the hands of the skilled stainless steel worker, but basic rules still apply; for instance, the same set of tools cannot be used for working ordinary mild steel as well as stainless. This is because ferrous material will be embedded into the stainless, leading to electrolytic action and corrosion at a later date. Once this form of corrosion has set in, it can be very difficult to arrest. For the same reason, stainless and mild steel stock material should not be kept in close proximity where they may come into contact. Again, note that proprietary test kits are available to check stainless steel for ferrous contamination.

The specification of the finish required on stainless steel can be a problem. It may be described in terms of polishing grit fineness or as *Roughness Average* (abbreviated to Ra). Roughness Average, also known as Centre Line Average, is a measure of the actual ridges and furrows on the metal surface and is measured in microns. Table 2.2 makes various comparisons about steel finishes.

Table 2.2 Stainless Steel Finishes

Description	Grit Number	Ra (microns)
Coarse brush	80	2.5
Fine brush	120	0.8
Dull mirror polish	180	0.4
Mirror polish (electropolish)	240	0.2
Bright mirror polish	320	0.1
Optical quality	500	0.05

In addition to these designations, there are various grades of bead blasting that may be applied to stainless steel isolator structures, such as shot peening, vacu-blasting, and vapour blasting.

A further process frequently used on stainless steel is electropolishing, in which a very thin layer of the surface of the materials is removed by electrolysis, resulting in a finish that may be described as satin polish. This process works on all surfaces of the object immersed in the electrolytic tank, but since it removes only a few microns of steel, the prefinishing must be of a very high standard at the start. A disadvantage, and one that characterises some forms of bead blasting, is that the weld lines appear, being of dissimilar material; however, electropolishing is now often specified for isolator finishes.

Clear specification of the finish required is important because the work is very labour-intensive and hence costly. As a general rule, a finish of 0.8 micron Ra is a good compromise on quality and cost.

Where extreme corrosion resistance is required, particularly in acidic conditions, then one of the Hastelloy range of alloys may be specified for all or part of an isolator structure. These alloys are composed of nickel and chromium with molybdenum, titanium, iron, and other additions, and are extremely expensive. They can be welded to ordinary stainless steel and can be used to form areas of particular resistance, without resorting to a complete Hastelloy structure.

Mild steel

Mild steel is a much cheaper and easier material to work with than stainless steel and does have application in some areas. Modern epoxy powder coating provides a tough finish on mild steel and, if the duty is relatively light, then this could be the material of choice. It is not likely to be acceptable in applications for licensed manufacture.

Air handling in isolators

Since it is the function of an isolator to provide a particular environment to suit the process in hand, the quality of the air or, indeed, of any other gas used to fill the isolator, will be of primary concern. This quality may be defined in terms of

- Filtration
- Pressure regime
- Flow regime
- Conditioning (temperature and humidity
- Composition (if not air)

Filtration

Where the isolator is to be used for sterile work, then a biological filter is required on the inlet air. Conversely, where a toxic material is to be contained, then a fine filter is needed on the exhaust air outlet. In practice, filters are fitted to both inlet and exhaust air, in both sterile and toxic applications, mainly to provide a definite barrier to the movement of particles through an unfiltered inlet or exhaust, but also to maintain pressure regimes.

The filters used are almost always HEPA filters. These are composed of a very fine-grained glass-fibre paper, arranged in various pleated forms to maximise the available surface area, either as panels or occasionally as canisters. HEPA filters are not simply sieves, as they arrest particles in the air passing through them by four main physical processes:

1. Straining or sieving effect. This takes place when the airborne particles are larger than the spaces between the filter fibres.
2. Impaction or inertial effect. This occurs when particles, though smaller than the fibre spaces, leave the airstream and impact the fibres through their inertia and are held by electrostatic forces. These electrostatic forces are strong, such that roughly 20 times the impact energy is required to remove a particle, once attached.
3. Interception. This occurs when particles essentially remain in the airstream, but the stream passes so close to a fibre that particles touch and thus attach to the fibre.
4. Brownian motion effect. The smaller particles in an airstream follow a rather random trajectory caused by Brownian motion, and this causes them eventually to leave the airstream and impact the filter fibres, where they are again held by electrostatic forces. Figure 2.1 illustrates these effects.

The efficiency of these processes varies with particle size. It is significant that the overall effect is to produce a minimum of efficiency at about 0.30 mm. This particle size is known as the Most Penetrating Particle Size (MPPS) and is a factor in filter testing (see Figure 2.2). The rate of airflow through these filters is directly proportional to the pressure drop across them; doubling the pressure will nominally double the flow rate. Broadly speaking, the flow through HEPA filters follows a form of Ohm's law.

The filter manufacturer will provide a graph to show this function, thus enabling the designer to size the filter to give the appropriate flow at a given pressure drop (Delta P). The rigorous designer may also take into account the fact that HEPA filters have maximum efficiency at a certain face velocity and size the filters to match this effect. As particles collect on the filter over time, the pressure drop required to maintain a given flow will increase. It is conventional to replace HEPA filters when the pressure drop increases to twice that of the new filter. This is the reason why pressure gauges may be specified across each HEPA filter in an isolator system. In practice, isolators

HEPA Filter Particle Capture Mechanisms

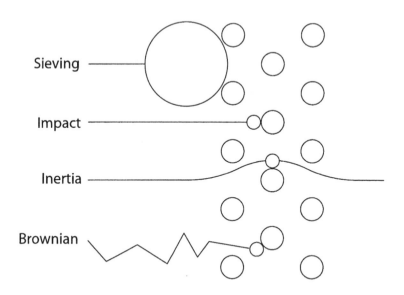

Figure 2.1 Particle Collection Efficiency. A graph showing the particle collection efficiency of the interception mechanisms and the diffusion mechanism in HEPA filters. The two effects combine to give the overall effect shown in the plot for total efficiency.

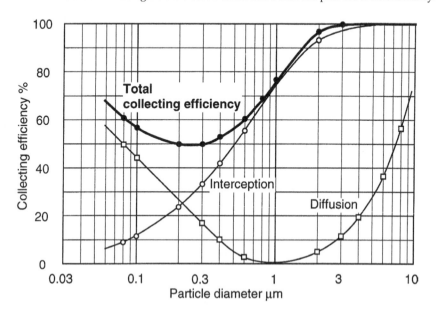

Figure 2.2 A Visualisation of the Various Mechanisms by Which HEPA Filters Remove Particles from the Air Flowing through Them.

tend to be operated in environments that are already clean; thus, filter life is very long, often in excess of three years.

The performance of these HEPA filters, in terms of their ability to arrest airborne particles, is measured as a function of the penetration rate of various sizes of particles. In the sodium flame test according to BS 3928, the particles are of sodium chloride, with a range of 0.02 to 2 mm in diameter and a median size of 0.65 mm, detected by means of a spectrophotometer. In the dispersed oil particulates (DOP) test, the particles were originally an aerosol of dioctyl-phthalate, but since this compound is now thought to be hazardous, a food-grade oil is used. The particles are detected with a sensitive instrument called a photometer. The all-important process of testing isolator HEPA filters is discussed further in Chapter 6.

The HEPA filter is usually defined as having an efficiency of 99.997 percent at the 0.30-mm particle size. In other words, the filter will pass 0.003 percent of a burden of particles that are 0.30 mm in diameter. To put this dimension into context, human hair is around 50 mm in diameter, fog droplets are about 10 mm across as are smaller pollens and fungal spores, while tobacco and oil smoke particles are around 0.10 mm in diameter.

Bacteria range from 0.30 to 20 mm on their longest axes, whilst viruses range from less than 0.01 mm up to 0.30 mm. Thus, the HEPA filter will arrest the majority of airborne bacteria and a large proportion of various smokes. They appear not to arrest viruses and this may be a significant issue; however, although viruses are small, they are nearly always bound to much larger particles or droplets, and the same considerations apply to agents such as the BSE prion. Figure 2.3a illustrates the size range of these various particles.

HEPA filters will give an excellent performance in terms of the particulate levels in most isolators, producing conditions much better than ISO Class 5, which is the norm for aseptic filling. In some cases, double inlet HEPA filters are specified as an added security where the process is critical or where continuous monitoring is not practical. Indeed, it should be noted that in the UK, the MHRA recommends the fitting of double inlet HEPA filters to aseptic isolators as standard practice. The MHRA feels that the filtration of inlet air for aseptic processing is so critical that the security of double inlet filters is well justified. Note that the logic is not to increase the overall efficiency of the inlet filters but rather to increase the security in the event of failure of one of the filters. By contrast, double exhaust filters are often fitted to isolators handling toxic materials for two reasons. The first is simply added security; the second is to allow for the safe changing of the primary filter, which may be loaded with a material of very low Occupational Exposure Limit (OEL). The primary exhaust HEPA filter is fitted within the isolator enclosure; the secondary is fitted outside. Thus, the primary may be safely demounted and bagged for disposal whilst inside the isolator containment. Meanwhile, the exhaust duct is protected by the secondary HEPA filter, which receives only a very small toxic burden and may be changed at the recommended service interval, with the minimum of safety precautions.

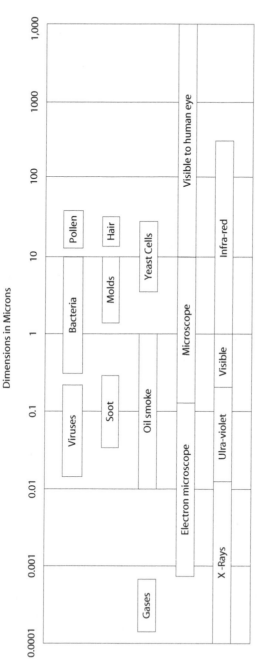

Figure 2.3a A Comparison of Various Particle Types and Their Size Ranges.

Figure 2.3b Double Safe Change Exhaust HEPA Filters Fitted to a Flexible Film Isolator. The primary filter is mounted directly to the base tray and is accessed from within the isolator by removing the lid, which, in this case, also carries the prefilter. The secondary filter is hung below the base tray and, thus, is accessed from outside the isolator. (Courtesy of Astec Microflow.)

Figure 2.3b shows the double exhaust HEPA filters on a flexible film, turbulent flow isolator.

A convention has developed recently in the U. K. in regard to toxic handling isolators, particularly those concerned with cytotoxic drug dispensing. The exhaust air may be returned to the laboratory if double exhaust filters are fitted; otherwise, it must be ducted to the atmosphere, via a negative-pressure duct if the isolator has only a single exhaust HEPA filter. The use of an exhaust duct has implications for the pressure regime, discussed below.

HEPA filters are quite adequate for most pharmaceutical and biotech applications, but the semiconductor industry demands higher standards for the process of wafer fabrication, where even submicron-sized particles can seriously affect the yield of the latest generations of microchip. Researchers talk in terms of ISO Class 2 or even ISO Class 1, which can be achieved only with the use of unmanned areas (i.e., isolators). This area of work creates many engineering problems due to the nature of the processes involved, but the use of ultra-low particulate air (ULPA) filters can at least provide very high-grade air indeed. These filters are rated as 99.9999 percent efficient at the 0.30-mm particle size.

To maximise the lifetime of HEPA and, even more so, ULPA filters, they should be fitted with prefilters that are easily changed or washed and replaced. Typical materials used here include various open cell foams, such as polyether and Bondina matting. These may have a performance of around 95 percent efficiency at 50 mm. Some isolator applications put a particular burden on the exhaust prefilter; for instance, research animal isolators need large and easily changed prefilters to cope with the high burden of airborne material produced by active rodents. Toxic isolators may also require special attention to the exhaust prefilter, where significant quantities of material can accumulate — for instance, in radiopharmaceutical applications.

As regards the mounting and sealing of HEPA filters, the conventional principles apply: they should be mounted as part of the duct rather than in the duct. There is no benefit in calling for a high-grade filter, only to leak contaminated air around a poor filter seal (see Figure 2.3c). The clamping mechanism for panel filters should be carefully designed to provide an even force around the seal, with no bending stress applied to the case of the filter. The clamping mechanism should also compress the seal by 50 percent of its thickness and no more. Ideally, the seal should be formed from a single piece of material, without a mitre joint at the corners. When properly designed and fitted, panel filters do not need either silicone rubber or silicone grease to form a leak-tight seal. Finally, the mounting of the HEPA filter should allow for reasonably easy change when required. Poor access to the filter makes correct fitting difficult and is thus likely to lead to leakage after the change has been made.

The overall design of an isolator air handling system will integrate the pressure regime, flow regime (see below), and any air conditioning (see below) with the specification for the filters. Since the HEPA filters usually represent the largest pressure drop in the system, they must be sized to match the required flow and the available pressure. For example, on a simple, single-fan, positive-pressure isolator exhausting to the atmosphere, the inlet filter is driven by the fan pressure, which might be 600 Pa, while the exhaust filter is driven only by the canopy pressure, which may be only 50 Pa. Thus, the exhaust filter will need to be much larger than the inlet to provide the same flow rate. The converse is true for a single-fan, negative-pressure isolator.

Generally speaking, larger isolators and those with unidirectional flow will be fitted with panel HEPA filters, whilst smaller glove isolators will commonly be fitted with canister HEPA filters. Panel filters may be scanned during *in situ* testing (see Chapter 6), but canister filters cannot be scanned and a volumetric test must be accepted. This means that whilst a canister filter may demonstrate an acceptable level of penetration under test, it may be that the leak is due to a single hole of microbiologically significant dimensions. For this reason, the airflow through canister filters should ideally be limited to around 150 m³/h in isolator applications. It will not be possible to scan some panel filters *in situ*, particularly exhaust filters, and again, a volumetric test must be accepted.

Some containment isolators will carry not only HEPA filters on the exhaust, but also activated carbon filters that will remove further

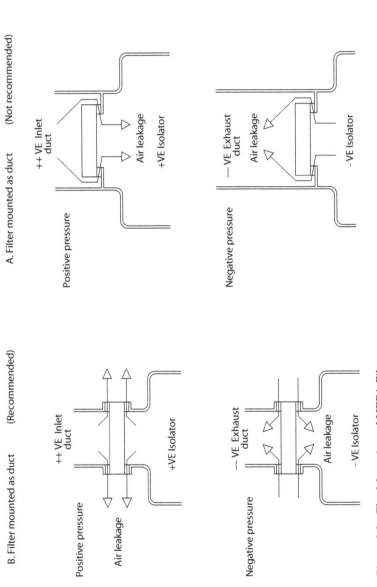

Figure 2.3c The Mounting of HEPA Filters

contaminants, particularly organics. These have a high pressure drop, and this must be accounted for in the design of the system. Whatever types of filter are fitted to an isolator, provision must be made to DOP test each individually, during validation and subsequent maintenance.

Note that this provision consists of three elements:

- A port to introduce the smoke challenge upstream of the filter
- A further port to sample the fully mixed challenge immediately ahead of the filter (e.g., in the plenum chamber)
- A final port to sample the airflow downstream of the filter (either by scanning or volumetrically)

The specification and testing of HEPA filters is not a simple matter, at the time of writing. Various national and international standards for the testing of filters after manufacture, and for subsequent *in situ* testing, are currently in draft form but likely to be promulgated shortly. Ideally, specific, current advice should be sought from the regulatory authority and the proposed filter manufacturer. For those in Europe, BS EN 1822–1:1998, High Efficiency Air Filters (HEPA and ULPA), is a recent standard, which uses scan tests but gives maximum values for both local and overall (volumetric) penetration. Figure 2.3d gives a table of the classification of HEPA and ULPA filters from BS EN 1822. In the UK, BS 3928 is still current at the time of writing and describes the sodium flame test, which is volumetric and does not provide for local measurement. Where *in situ* leak testing is required, BS 5295, Environmental Cleanliness in Enclosed Spaces, provides methods, while BS PD 6609–2000 gives general guidelines. Eventually, specification and testing should be harmonised under BS EN ISO 14644–3, Metrology and Test Methods; however, this is currently in draft form only.

Figure 2.3d A Classification of HEPA and ULPA Filters from BS EN 1822-1: 1998

Filter Class	Overall Value		Local Value	
	% Efficiency	% Penetration	% Efficiency	% Penetration
H10	85	15	—	—
H11	95	5	—	—
H12	95.5	0.5	—	—
H13	99.95	0.05	99.75	0.25
H14	99.995	0.005	99.975	0.025
U15	99.9995	0.0005	99.9975	0.0025
U16	99.99995	0.00005	99.99975	0.00025
U17	99.999995	0.000005	99.9999	0.0001

As a generalisation, the following factors need to be addressed when specifying HEPA filters for isolators:

- The volume flow rate (or maximum intended face velocity)
- The filter case size — length, width, and depth
- The filter case material — normally extruded aluminium (not wood, plywood, MDF, hardboard, etc.)
- The pressure drop across the new filter
- The seal type, material, and location (clean side, dirty side, or both)
- Manufacturer statement of the MPPS at the intended flow rate
- The intended *in situ* leak test method

Figure. 2.4 Standards for the Testing of HEPA Filters

BS EN 1822 Parts 1–5. 1998–2000	High Efficiency air filers (HEPA and ULPA)	Published
BS EN ISO 14644 Part 3	Cleanrooms and Associated controlled Environments – Metrology and Test Methods	DIS (Draft International Standard) Publication expected by 2005
PD6609–2000	Environmental Cleanliness in Enclosed Spaces – Guide to test methods	An interim document pending the pubication of BS EN ISO 14644–3

Pressure regimes

Isolators are normally maintained at a specific pressure differential with respect to the isolator room and to other parts of the isolator system. In sterile or other clean operations, such as semiconductors, the isolator will be held at a positive pressure with respect to the room, so that any leakage, small or large, will be outward from the isolator, thus maintaining clean conditions. In toxic containment applications, the isolator will be held at negative pressure, so that any leakage will be inward, thus protecting the operators. Neutral pressure is an option, given a two-fan system, but one that is rarely employed.

Operating differential pressures range from 30 to 200 Pa (3–20 mm of water gauge), but the range 30–70 Pa gives better operator comfort (see below) and less likelihood of leakage. Systems that have several isolators interconnected, as with pharmaceutical filling lines, can be designed to have a pressure cascade with the most critical unit having the highest differential, falling through the system via the less critical isolators or lockchambers. Where the product is toxic and yet must be maintained in sterile conditions, as with cytotoxic drugs, an interesting dilemma develops, as discussed in Chapter 5.

Another problem that may need to be addressed when considering pressure issues and, indeed, flow issues, is that of exhaust ducts. It is quite common to take the exhaust from a toxic isolator to the atmosphere through a duct and, in the case of sterile isolators, a duct may be needed to remove exhaust-sterilising gas. Safety regulations require ducts carrying hazardous gases to be under negative pressure throughout their length, so that any leakage will be inward. This then requires the addition of a fan at the atmospheric end of the duct, capable of overcoming the pressure drop of the duct, at the working airflow rate.

This sounds like a fairly simple engineering exercise, but there are some difficult issues to address, because the addition of the remote fan creates a double-fan situation. It is worth noting this particular problem here. Where an isolator is fitted with a single fan, be it positive pressure or negative pressure, exhausting to the atmosphere, the isolator pressure and flow will be stable for any given fan speed setting. In marked contrast, if one fan is fitted to the inlet of the isolator and another to the exhaust, an unstable situation is created — one that is not easy to control. This is inherent in the nature of the centrifugal fans used on isolators. Where one centrifugal fan is balanced against another, very slight changes in the speed of one fan make dramatic changes in the dynamics of the system. These changes can be caused by minor factors, such as the temperature or pressure of the atmosphere, or small changes in the electrical control and performance of the fans.

If the two fans in a double-fan system are very close to the isolator, or if the ducts between the fans and the isolator are very wide with little pressure drop, then the isolator control system can cope normally, whether pressure-governed or flow-governed.

If, on the other hand, there is a long and narrow duct between the isolator and the exhaust fan, with a large pressure drop, the system may be unstable to the extent of being uncontrollable. Worse still, the time constants of such a system may be such that all may appear well on the day of commissioning, only to alter significantly some days later. One solution to this problem is to decouple the isolator from the duct by use of an air break or "thimble," but this has further safety implications. Here are some suggestions:

- Avoid double-fan situations if possible.
- If a remote exhaust fan is required, make the duct as short, as straight, and as wide as possible.
- Use something other than a centrifugal fan to drive the incoming air, such as compressed air from a conventional 7 bar supply. This high pressure will need to be dropped through a suitable pressure-reducing valve.

There are implications for the building system when considering the installation of an isolator requiring a duct to the atmosphere, and these should be addressed at the earliest possible time in the project. The next section, on flow regimes, may also have some impact on the building system.

As a final comment on pressure regimes, designers may need to consider fitting some form of pressure-relief device to the isolator. If, for example, the isolator is fitted with a supply of compressed air, the valves controlling the supply may fail, or may leak over time, thus raising the pressure of the isolator to potentially disastrous pressures. In the case of toxic containment isolators, the pressure relief device must exhaust to a safe place, such as the exhaust duct. ISO DIS 14644–7 describes a suitable oil-filled device for use particularly with inert atmosphere isolators.

Flow regimes

The pattern of the airflow through an isolator may take one of three forms: turbulent flow, unidirectional flow, or a combination of the two.

Turbulent flow

Turbulent flow is the simplest form of flow regime: air is introduced at one point and exhausted to the atmosphere (or at least to atmospheric pressure) from another. Conventionally, the flow rate through a turbulent isolator will give an air-change rate of between 10/h and 100/h, with most isolators set at around 25/h. Although this regime is simple, some thought needs to be given to the disposition and the nature of the air input and exhaust. The ideal flow pattern gives full purging of the entire volume of the isolator, with no standing vortices or dead volumes where airborne particles might accumulate and then precipitate. The input should be as far from the exhaust as practical. Ideally, the input should incorporate some form of distribution device to spread the inlet air across the isolator volume. The position of the exhaust may be dictated to some extent by the need to fit a changeable prefilter, easily accessed by the operator. It is not entirely simple to demonstrate good purging in a turbulent isolator, but introducing DOP smoke into the inlet airstream can indicate the flow pattern quite well, and the time taken to remove the smoke will indicate the degree of mixing.

Since the volume of air flowing through the isolator is quite small, perhaps 30 m^3/h for a standard four-glove isolator, only small fans of about 200 W are needed to move air through the system, and the exhaust is often directly back to the isolator room. Consideration should be given, however, to the question of breach velocity, where turbulent flow is used in negative-pressure containment isolators. The breach velocity is the velocity of air passing through any inadvertent break in the isolator wall, and this is normally taken to mean the loss of a glove from its cuff ring. This velocity should be at least 0.70 m/sec to prevent the loss of airborne material to the outside. If the cuff aperture is 100 mm in diameter, this means an inflow of air at the rate of 20 m^3/h. Clearly, the exhaust fan must be capable of providing this flow as a minimum. Figure 2.5 shows the ventilation system of a typical turbulent flow isolator.

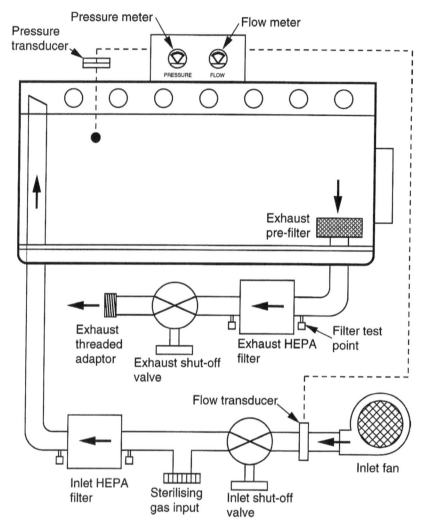

Figure 2.5 The Components of a Typical Small, Turbulent Flow, Positive-Pressure Isolator.

Unidirectional flow

The original term used here was *laminar flow,* but the term *unidirectional flow* is now more generally accepted because laminar flow may contain velocity gradients, whilst unidirectional flow is taken to have not only uniform direction but also uniform velocity. This regime was developed in various types of biological safety cabinets and in cleanrooms. The flow is usually vertically downward, although horizontal flow is used in some cases. The airflow is set to emerge from a panel HEPA filter at a velocity of around half a metre per second, the regulatory figure being given as 0.45 m/sec ± 20%. The true origins of this figure are somewhat obscure but seem to be empirical and linked to the original design of horizontal laminar flow cabinets. In practice, velocities down to 0.05 m/sec appear to give good results in isolators (see below).

Under these conditions, the airflow is unidirectional, with all the flow in one direction at one speed, which means that the flow will not entrain air from outside the flow. Thus, for instance, particles shed by an operator in a cleanroom with unidirectional downflow air will be drawn immediately to the exhaust at the floor and not allowed to rise up to the level of the work surface. If conditions were turbulent, then these particles would circulate with the air and be presented at the work surface.

The application of unidirectional flow was a major step forward in contamination control and so is now more or less obligatory in critical processes, such as the aseptic filling of pharmaceutical products. As a result, unidirectional flow is often specified in isolators, but this is not entirely logical. The concept was mainly developed to reduce the entrainment of particles shed by even fully suited operators; this problem does not arise in the isolator, where operators are, of course, physically and biologically excluded. The indicated view of the MHRA is that unidirectional flow is desirable, but not obligatory (Bill 1996).

Since isolator chambers are only about 750 mm high, the only way to approach true unidirectional downflow is to have a perforated floor. This then leads to other problems with spillages and leakages, so that the conventional way to exhaust air from a unidirectional flow isolator is via gutters along the front and back edges. In this case, laminarity can only extend perhaps halfway down the isolator before the flow breaks up to reach the exhausts, in which case the equipment near the floor of the isolator is bathed in turbulent flow. Furthermore, the sleeves or half-suits, and very often the equipment in an isolator, constitute a large fraction of the isolator volume and so readily cause turbulence.

Unidirectional flow also demands much more complex engineering in an isolator because the flow rates are much higher. For example, a four-glove isolator measuring 2000 mm long and 650 mm wide might have a flow rate of 50 m^3/h in a turbulent regime. By contrast, in a unidirectional regime, it would have a flow rate of over 2000 m^3/h. We then have to decide how to handle this volume of air:

- Total loss. If the application is toxic, then the flow should be exhausted to the atmosphere, in which case the isolator room heating and ventilation system must cope with this major loss.
- Return to the isolator room. The fan size needed to cope with this flow is quite appreciable and will dump several kilowatts of heat into the room. Once again, the heating, ventilation, and air conditioning (HVAC) system must be sized accordingly. The noise generated by the exhaust of this airflow may also create a problem.

- Recirculation. Most of the air leaving the isolator is returned to the top plenum chamber for recycling. A proportion, perhaps 10 percent, is exhausted and a corresponding volume of fresh air is introduced to the system — the makeup air is usually taken from the isolator room and may be exhausted to the atmosphere or returned to the room. This solves the problems of the first two options, but the engineering becomes complex: large return ducts must be provided to transfer the air from the base of the isolator to the top plenum. It is desirable to keep all air duct velocities below about 5 m/sec to keep pressure losses low (see below) and to reduce noise. Given a flow rate of 2000 m³/h, as in the standard four-glove isolator mentioned, and a duct velocity of 5 m/sec, we have to incorporate a return duct in the isolator with a diameter of almost 400 mm. As previously mentioned, the fan power needed to move this volume of air will amount to several kilowatts, which will quite quickly heat up the air in the recirculating system. Thus, air cooling will have to be incorporated to maintain the usual 20°C atmosphere inside the isolator (see below).

Having noted the disadvantages of unidirectional airflow in isolators, the significant advantage of the regime is the speed with which particles are purged from the critical working area. This is particularly useful in situations where both asepsis and containment are required, such as the processing of cytotoxic drugs. The problems associated with this type of work are discussed in more detail in Chapter 5, but briefly, negative pressure regime is often applied to cytotoxic isolators to protect the operators. Any leakage, however, may draw contamination into the isolator and put the patient at risk. If a unidirectional airflow regime is used, such particles are removed from the critical work area within seconds. Under a turbulent airflow regime, the particles may remain in circulation for minutes, thus increasing the chance of product contamination. For this reason, hospital pharmacy isolators used for cytotoxic work in the UK often use the unidirectional airflow regime. (Neiger, J.S., Negative Isolation, *Cleanroom Technology*, 2001; April: 24–25.)

In the final analysis, the chosen airflow regime must reflect the balance between factors such as the sensitivity of the process in the isolator, the air quality of the isolator room, and the leak-tightness ("arimosis") of the isolator.

Figure 2.6 is a schematic drawing of the airflow through a unidirectional airflow isolator.

Semiunidirectional flow

Whilst this may sound like a contradiction in terms, this semiunidirectional flow has found use in some isolators, such as those used for aseptic hospital pharmacy work. In this design, just a part of the roof of the isolator is made up of a panel HEPA filter, which generates a region of unidirectional flow

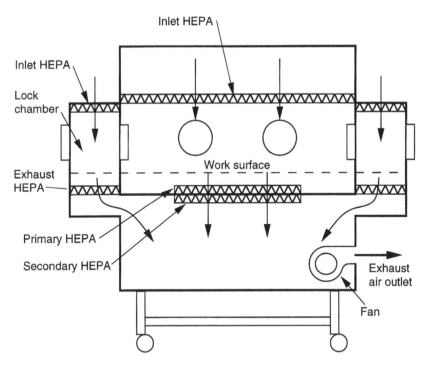

Figure 2.6 Airflow Schematic. A schematic of airflow through a unidirectional down-flow isolator of the design often used in hospital pharmacies for cytotoxic dispensing. Note the integration in the design of the lockchambers used for transfer.

within the main body of the isolator. Curtains or plates are fitted to contain the flow down to perhaps 400 mm from the isolator floor, and air then spills out to give turbulent flow in the rest of the isolator. Thus, small-scale critical operations can be carried out in the region of unidirectional flow, while less critical work goes on in the main body of the unit. All of this can be engineered without recourse to the complexities of full unidirectional downflow. Figure 2.7 shows the airflow pattern of a semiunidirectional isolator.

Although unidirectional airflow regime isolators are complex and expensive, and the benefits may be perceived rather than actual, it seems likely that they will still be used in aseptic pharmaceutical operations. Change is only likely to occur in the wake of extensive data provision, which in turn means that a fairly comprehensive research programme is required. Possibly, such a programme could be coordinated by one of the learned bodies, such as the Parenteral Society (UK) or the PDA (U.S.), and the work shared among a number of pharmaceutical companies.

Air conditioning

A further aspect of air handling concerns the temperature and humidity of the isolator air, and this has already been touched on in the previous section.

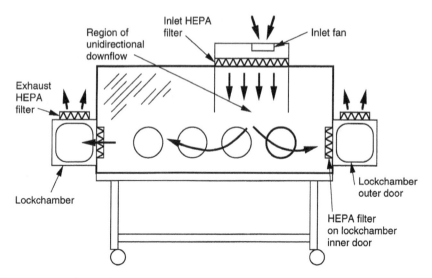

Figure 2.7 Airflow Schematic. A schematic of airflow through a semiunidirectional downflow isolator. This type of flow regime is often used for aseptic dispensing in hospital pharmacies.

In simple isolators with small fans, returning exhaust air to the room, and where little or no heat is generated by the process, conditions in the isolator will be pretty much the same as those in the room.

If, on the other hand, the isolator handles large volumes of air, has recirculation, or houses equipment that generates heat, then careful consideration will have to be given to air cooling. This may be self-contained, with refrigeration systems on board the isolator, or it may make use of a cooled water supply; in either case, suitable cooling coils must be incorporated into the air system.

Humidity may also be an issue in some applications. If low humidity is required, then there are several options:

- Refrigerant drying. Two systems with automatic changeover for defrosting may be required.
- Silica-gel bed. Two systems are often used with automatic changeover for regeneration.
- Rotary (lithium chloride) dryer (e.g., Munters Rotaire).
- Nitrogen. In some small applications, dry nitrogen may be used from a cylinder.

If high humidity is needed for the process, then some form of humidifier must be fitted and the exhaust from the isolator arranged accordingly.

Temperature and humidity are, of course, bound up with one another, and the design of the isolator conditioning system may need to take this into account. Isolators are generally good heat exchangers. It is not easy to run them at temperatures very different from the room that houses them. If this is required, then insulation of the isolator body will be needed. The control of isolators with air handling systems that incorporate recirculation and air conditioning becomes more complex, and very often a PLC will be fitted, as discussed in Chapter 4. If clean-in-place/sterilise-in-place (CIP/SIP) systems are incorporated into the isolator, then the piping and instrumentation diagram (P&ID) will need some very careful design to meet the criteria of each subsystem. Figure 2.8 shows the P&ID for a robotic transfer isolator for use in a large-scale aseptic filling and freeze-drying application; this isolator must be cleaned and resterilised in the brief period between vial transfers, giving an interesting challenge to the design engineers involved.

Calculations for isolator air handling systems

The purpose of this book is not to describe the detailed engineering of isolators, but there are a number of simple formulae that may be of use to those examining aspects of isolator air handling.

- Air velocity in ducts

$$v(m/sec) = \frac{flow\ (m^3 sec)}{cross\text{-}sectional\ area\ (m^2)}$$

- Pressure drop of airflow in circular ducts

$$Pressure\ loss/metre\ (Pa) = \frac{91.9 \times v^{152}}{diameter^{1.269}\ (mm)}$$

- Velocity head
  - Velocity head $\times 0.5 \times \rho v^2$ (m/sec)
  - In practise, value used $= 0.6 \times v^2$ (m/sec)
- Pressure drop in elbows
  - Pressure drop (Pa) through 90° sharp elbow = velocity head $\times 1.25$
  - Pressure drop (Pa) through 90° lazy elbow = velocity head $\times 0.25$
- Heat loss to airflow

$$airflow\ (m^3/h) = \frac{kw \times 3420}{temperature\ rise\ (°C) \times 1.02}$$

The characteristics of HEPA filters are usually described by a graph of flow rate against pressure drop, provided by the manufacturers. Similarly, the performance of centrifugal fans is usually described by a graph of flow versus pressure developed, and supplied by the manufacturers.

Airflow modelling in cleanrooms has become very sophisticated recently, and such computer-based modelling may well be of use in some aspects of isolator design in the future. Several specialist companies can provide services in this area, and isolator designers might wish to contact them.

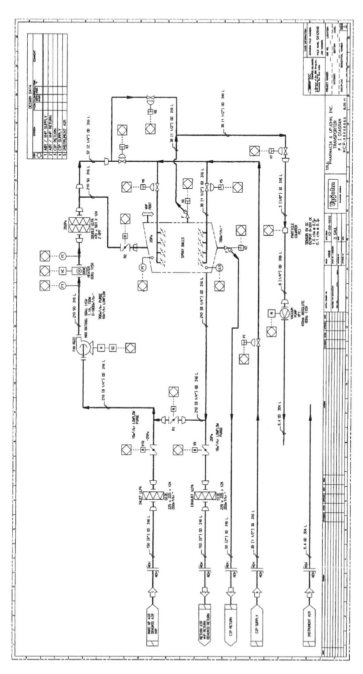

Figure 2.8 The P&ID of a Special Isolator Fitted with a CIP/SIP System and Integrated Instrumentation. This shows the complexity that quickly develops as more systems are added to an isolator. (Courtesy of Astec Microflow.)

Handling the work in isolators

There are four basic methods, or *Access Devices* as they are termed in some guidelines, by which the operator can access the work to be carried out inside isolators.

Gloves

Gloves and sleeves are the simplest method of handling and are the method of choice provided that the work can be accessed. The maximum arm length of most operators is 700 mm, with a realistic limit to the working radius of 500 mm. The general weightlifting limit for working in sleeves is reckoned to be about 5 kg, and the absolute limit is 10 kg. A simple test to judge if a specific process can be handled when working in sleeves is to try out the process while seated at a bench. If the work can be handled reasonably comfortably in this way, without moving from the seated position, then it can probably be carried out in a glove isolator.

Gloves in isolators may take several forms. Gauntlets are the simplest type, used on early isolators and now used often on containment isolators. The gauntlet incorporates a glove and a sleeve in one piece, which is sealed firmly to the front window of the isolator. They may be made of various materials, such as rubber latex, butyl rubber, nitrile rubber, or Hypalon™, the choice being dependent on the materials and solvents to be handled and also on any sterilising agent to be used. Gauntlets represent the most robust method of handling, but not the most comfortable, whilst the short types give very restricted reach. Changing the gauntlets usually involves a break of containment, unless one of the sophisticated, nuclear glovebox shoulder rings is specified.

Many glove isolators will now have separate sleeves that are sealed onto shoulder rings set in the front window of the isolator. These sleeves have cuff rings to carry the gloves, which may be changeable by various propri-etary devices. The logic here is that the gloves are the most vulnerable part of most isolator systems and should be easily changeable, preferably without breaking containment. The second most vulnerable parts are the sleeves and so these, too, should be changeable, though not necessarily without breaking containment. A popular form of construction for isolator sleeves has two layers: the first is a proofed woven fabric that gives a comfortable cloth surface on the operator side, and the second is natural PVC that gives a smooth sanitary surface on the isolator side. These two materials are laid together and fabricated into a sleeve, usually by RF welding. The two remain in close contact unless either layer is punctured, in which case they separate readily. Thus, these sleeves function well and are self-indicating of leaks. Figure 2.9 shows double-layer sleeves, whilst Figure 2.10 shows PVC sleeves welded as part of the canopy structure.

The choice of gloves to be used on the isolator is not altogether simple. The thinner and lighter the glove, the better the feel and dexterity, but the

Figure 2.9 Typical Double-Layer Sleeves and Quick-Change Gloves Fitted to a Hospital Pharmacy Isolator. (Courtesy of Astec Microflow.)

greater the risk of tearing or puncturing. It was noted by Thomas and Fenton-May (1994) that solutions of cytotoxic drugs pass through the wall of almost all glove materials in a short period of time. In this case, the only safe response is to change the gloves on a regular basis, before the contamination has time to diffuse through to the user.

Large-size gloves allow the user to leave the gloves without everting them, ready for the next user, but may be less dextrous in use. Some operators

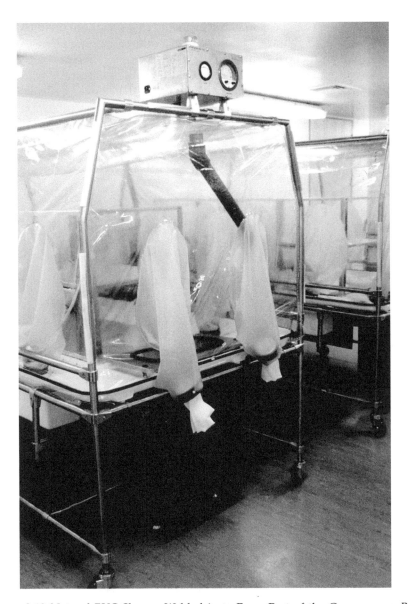

Figure 2.10 Natural PVC Sleeves Welded in to Form Part of the Canopy on a Bio-medical Isolator. (Courtesy of Astec Microflow.)

fit thin, latex surgeons' gloves, in which case they are changed for each batch of work. Others may use cotton or silk undergloves worn inside the isolator gloves for comfort during extended work periods. The *Yellow Guide* (Lee and Midcalf 1994) suggests that operators undertake a hand disinfection before entering isolators for sterile applications. In general, it is a good plan for operators to test a number of gloves and select the best overall compromise. It is very useful to have a glove-change system that will accept a variety of glove types, which can be changed reasonably easily without breaking containment. Figure 2.11 shows a telescopic system for changing gloves, in this case on a half-suit. Gloves should be arranged in pairs wherever possible and the use of ambidextrous gloves should be avoided, as these tend to be uncomfortable in use. The *Yellow Guide* requires that gloves be leak tested every day, a process discussed in Chapter 6.

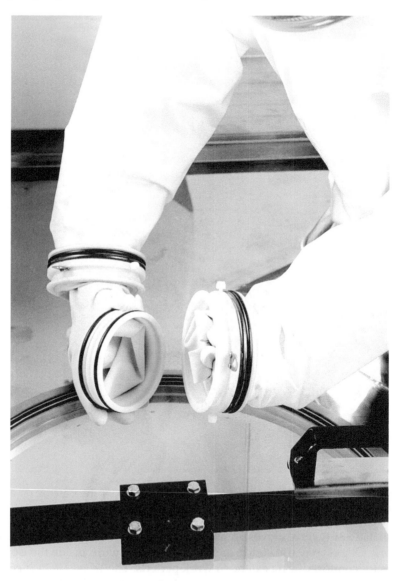

Figure 2.11 Telescopic Cuff Rings Allowing Easy Glove Change without Break of Containment. A new glove and ring are about to be fitted by pressing out the old glove and ring, which are to the right of the picture. (Courtesy of Astec Microflow.)

Half-suits

Where access to a larger area of work is required, or where heavier items must be moved within an isolator, the half-suit is used. As the name suggests, it is a sealed suit, complete with its own air supply for the operator, but one that covers the user only from the waist upward. It is mounted on an oval flange in the isolator base tray, often on an angled plinth to aid entry and exit. Thus, the user can work inside the isolator, much as if working at a conventional laboratory bench, but still be physically separated from the contents of the isolator. The maximum reach when working in a half-suit is around 1200 mm, with a weight limit of about 15 kg.

There is often some resistance to this concept, both on the part of engineers involved with the design of new isolator systems and the users who will have to work under these conditions. This is perhaps a very natural response, but the half-suit is, in fact, a very practical device and widely used in isolation technology. From an engineering point of view, it provides much greater handling capacity than sleeves and gloves; without it, more automation may be required to carry out the work. Alternatively, the isolator might need to be decommissioned and opened for such work as filling machine maintenance, without the use of a half-suit. From the user's point of view, the great benefit is the ability to work in a fully sterile or fully contained environment without the need for cumbersome protective clothing, such as that required in a cleanroom. The user can enter and leave the suit very quickly to answer the telephone or go for breaks, without compromising the isolator. Users will probably wear a lab coat and mobcap but, other than this, there is little restriction involved when using half-suit isolators. Once operators have used the suit one or two times, they are normally quite happy to continue working with them. The suit is less comfortable in positive-pressure applications because it tends to compress gently onto the operator, but this is no real hardship.

The half-suit has its own air supply, usually feeding to both the face region and the cuffs, to give a steady flow of air over the whole body. Very often, this supply is HEPA filtered as an added precaution, which makes for pleasant air quality. It is important to note that half-suits should not be entered, even briefly, without the air supply running. If the isolator has recently been gas sanitised, then the air supply should be run for at least 30 minutes before use, to remove traces of gas that will have diffused through the suit. The users may work standing, or they may use a high stool or chair and work seated.

Materials similar to those used for sleeves are popular for half-suit construction, these being tough but flexible, and resistant to the action of cleaning and sanitising agents. Half-suits need to be light in weight to be comfortable, while the air supply should be effective but unobtrusive. The skirt of the suit needs to be full enough to allow the operator to reach across the isolator and to turn easily in either direction, but not so full as to leave folds of material inaccessible to sanitising gas. Some half-suits have a quilted two-layer construction. The interspace between the layers is used as a duct to take air from the input hose to the neck and cuffs of the suit. This is a good system, since this pathway remains open whichever way the operator moves or leans. A further advantage of the quilted suit is that, to some extent, it reduces the effect of the suit pressing onto the operator in positive-pressure isolators. The disadvantage of the double-layer suit is the overall bulkiness and increased weight. Half-suits need to be suspended when not in use, to present the suit ready for the operator to enter easily and to expose all the surfaces of the suit during gassing. Probably the best way to suspend the suit is from the cuffs, with the arms raised in the "don't shoot" position. The suspension system should disconnect easily once the operator is in place and the use of elastic suspension cords, commonly used in the past, should be avoided. It is important that the faceplate be optically clear and undistorted if the user is to work for extended periods. Some suits have incorporated intercom systems, which makes communication with other suit users and with those outside the isolator much easier. In some cases, a fold-down platform is fitted to the isolator support structure to give shorter half-suit operators a raised standing area. A further consideration with the use of half-suits and, indeed, for isolator sleeves and gloves, may be interlocking, in relation to moving machinery. This is discussed further in Chapter 4. Figure 2.12 shows the author using a half-suit in a typical flexible film isolator, while Figure 2.13 shows a half-suit from within the isolator.

Full suits

Full suits have found some limited application in isolation technology where very extensive work access is required. One particular design uses an adaptation of the RTP (see Chapter 3). Here, the space between the container lid and the port door is effectively enlarged to accommodate an operator. The container lid becomes a lower half-suit and the port door becomes an upper half-suit. In this way, the operator can be "posted" into the isolator. An umbilical line supplies breathing air and communication for the operator whilst at work.

Robotics and automation

One obvious route for handling in isolators is to automate the process as much as possible, or to install robots to carry out all or part of the work. Robots have certainly been used in isolators to perform repetitive work, such as sterility testing by direct inoculation on large numbers of samples. This has been tolerably successful, but the majority of isolators are still run by human operators. One problem with robots is the question of sterilisation, where the process is sterile, and this has been addressed by covering the robot with a PVC gaiter, as used in the nuclear industry on master-slave manipulators.

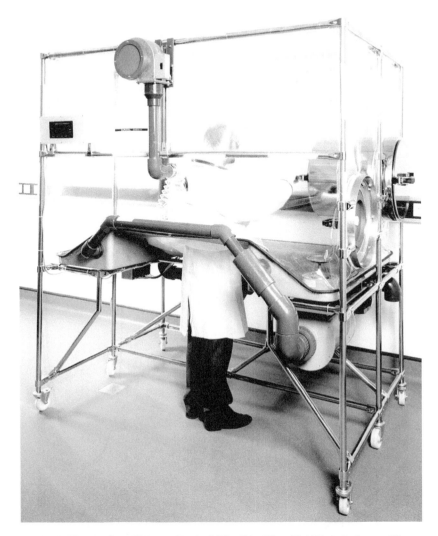

Figure 2.12 The Author Using a Typical Flexible Film Half-Suit Isolator. (Courtesy of Glaxo Manufacturing Services Ltd.)

Figure 2.13 A Half-Suit in Use Viewed through the Open Transfer Port. Note the communication system fitted to this suit.

chapter three

A review of transfer methods

The loading of materials into isolators and the subsequent removal of products and waste, without breaking containment, is probably the single most taxing issue in isolation technology. A large number of devices have been produced for this purpose, and the choice of the most appropriate will depend mainly on the nature of the work to be carried out. Some of the devices are very simple and consequently allow for mixing of the isolator and room atmospheres to some degree. Others are more sophisticated and allow no intermixing whatsoever. The *Yellow Guide* (Lee and Midcalf 1994) considers transfer to be the principal factor governing the quality of the isolator room environment, an important issue discussed further in Chapter 5.

Before proceeding, it may be of use to define some basic terms to be used in this section. The term *transfer hatch* is often applied, in conversation, to methods of isolator input and output, particularly referring to lockchambers, but also to other types. This is a vague term and it is proposed that it be abandoned in favour of one blanket term to refer to all methods of input/output, with clearly defined names for each type of method. The terms chosen for this book are intended to convey clear and logical meaning, but they are at variance with some existing definitions (see the Glossary). The suggested blanket term is *transfer port*. The terms used for each type of transfer port are given in the rest of this section; however, mention should perhaps be made at this stage of the classification of transfer ports given in the *Yellow Guide* (Lee and Midcalf 1994). This system grades transfer ports on a scale from A through F, according to the quality of transfer, in terms of the intermixing of the isolator environment with the room environment, given by each type of port in operation. Thus, simple doors on isolators are classified "A," since there is free exchange of air between the room and the isolator interior when the door is used. In the midrange of the scale, there are various types of lockchambers with different degrees of air exchange in use. Toward the end of the scale, docking devices and direct connection of the isolator to equipment, such as an autoclave, are classified "E," since they allow little intermixing. The principal purpose of the scale is to help in defining the room environment required to house any particular isolator, as described in Chapter 5. The reader may find reference to the scale in product literature where, for instance, a particular isolator may be described as having type D lockchambers fitted. New devices are under constant development, designed to address particular problems, but the main types are described below.

Simple doors

Simple doors are used to load isolators with equipment or quantities of materials before closing the door to seal the isolator before starting work. Clearly, they allow more or less free exchange of air between the isolator and the room, so a toxic isolator must be decontaminated before a door is opened and an aseptic one must be sterilised before work starts. In its most basic form, the door may be a flexible PVC cap, placed over a stainless steel ring in the isolator wall; this is commonly called a "jampot" cover.

Most simple doors consist of a sheet of clear acrylic or polycarbonate material, hinged along one edge and closed, by a latch on the opposite edge, onto a flexible seal. This is workable in small sizes, but the pressure on the seal is applied in only three, or perhaps four, places, which distorts the door. If the door is large, the distortion may lift it away from the seal in places, giving leakage: pulling down the hinges and latch harder only distorts the door further and increases leakage. Large doors will, therefore, be stiffened and pressed to the seal in a number of places. One form of a simple door on the market overcomes these problems by applying a high closing force at the centre of a circular door, using a pressure bar and latch.

If the isolator is to be used for pharmaceutical applications, in particular licensed manufacture, then the door seals, as with all other isolator seals, should be made with an accepted material (e.g., Viton™ or silicone rubber).

Among simple doors, we can also include other opening devices, such as airtight zips, which are occasionally used to gain access to isolators. Figure 3.1 shows a circular simple door on a flexible film glove isolator.

Lockchambers

Various types of lockchambers are fitted to isolators, so it may be useful to first define the term: A lockchamber is a subsection of an isolator, with a sealable door communicating with the main volume of the isolator and a further sealable door communicating with the exterior. It is intended that only one of these doors may be open at any one time.

In its simplest form, the lockchamber has no airflow and is used merely as an attenuating device between the isolator and the exterior. Figure 3.2 shows a simple lockchamber with no air purging. Most types of lockchamber will have some form of air purging, as follows:

Figure 3.1 A Simple Door Fitted to a Flexible Film Isolator. In this case, the door is circular clear Perspex and fitted with a single latch. This is the isolator previously seen in Figure 1.3. (Courtesy of Astec Microflow.)

Figure 3.2 A Very Simple Circular Lockchamber Mounted on a Negative-Pressure Flexible Film Glove Isolator. This lockchamber has no form of air purging and relies on manual cleaning in a mildly toxic production application.

1. Active. The chamber has its own dedicated fan and HEPA filters to provide airflow. Pressure may be positive or negative but will usually be roughly midway between the isolator pressure and the exterior, to give a cascade effect. The lockchamber is, in effect, a miniature isolator and may even be provided with gloves to carry out loading or decontamination operations within the chamber. Indeed, a complete isolator may be used to form a lockchamber used to load and unload a larger working isolator; this is often the case in sterility-testing operations.
2. Passive. The lockchamber is fitted with HEPA filters arranged in such a way that all or part of the exhaust from the main isolator passes through the lockchamber to produce a strong purging effect. This arrangement, in various forms, is commonly used on hospital pharmacy isolators where speed and simplicity of transfer are prime requirements.

Both active- and passive-flow lockchambers can be fitted with connections to allow gas-phase sterilisation for sterile transfer. Some are fitted with spray ports for wet chemical sanitisation, as used in research animal housing. Some instrumentation, perhaps in the form of a Magnehelic™ pressure gauge, may be fitted to lockchambers.

The booklet *Isolators for Pharmaceutical Applications* (Lee and Midcalf 1994) describes various classes of lockchambers as a part of the overall classification of transfer devices. The classification has been further described by John Neiger (1997) ; some useful schematic sketches of the various transfer devices in operation are included in the paper, and these are reproduced in Figure 3.3. Note that the transfer devices in the classes A, B, C1, and C2 do not come within the strict definition of isolation discussed in Chapter 1 because there is, technically speaking, a direct pathway between the critical zone and the room atmosphere. Lockchambers of the classes C1 and particularly C2 have, however, been used with good reported results on hospital pharmacy isolators.

Figure 3.4 shows a passive lockchamber fitted to a positive-pressure flexible film isolator for hospital pharmacy use. This is a type D lockchamber with HEPA-filtered air purging via the filter mounted on the inner door and the filter on the roof of the chamber.

Figure 3.5 is a lockchamber fitted with an electromagnetic interlock system, which prevents the opening of both doors at the same time. It can also lock both doors or open both, by use of the key provided. A HEPA filter will be fitted to the aperture on the roof of the chamber, showing that this is a type D lockchamber.

Figure 3.6 shows a pull-out drawer fitted into a lockchamber on a rigid wall, unidirectional downflow isolator. This drawer makes the removal of items from the lockchamber much easier for the isolator operator, though it creates some cleaning problems. The handle-like fitting on the door is a part of the mechanical interlock applied to this lockchamber.

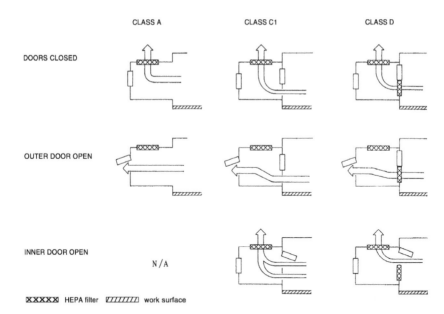

Figure 3.3a Various Types of Lockchambers. A schematic illustration of the action of A, B, C1, C2, and D class lockchambers with both doors closed, outer door open, and inner door open. (Reprinted from J. Neiger, 1997.)

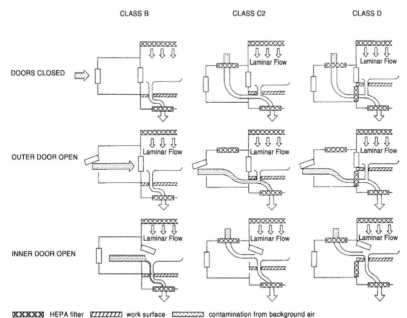

Figure 3.3b

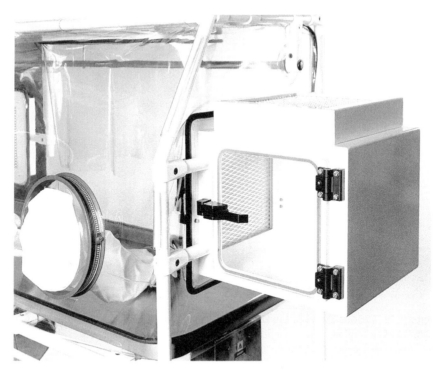

Figure 3.4 A Passive Lockchamber Fitted to a Positive-Pressure Flexible Film Isolator for Hospital Pharmacy Use. This is a Class D lockchamber with HEPA-filtered air purging via the filter mounted on the inner door and the filter on the roof of the chamber. (Courtesy of Astec Microflow.)

Product passout ports

Product passout ports are a type of "bagging port," which provides for semicontinuous output of relatively small items, such as prepared syringes, TPN bags, or waste material. The principle is to provide a quantity of layflat polyethylene tubing reefed onto a cartridge, so that an item can be passed out of the isolator into this polyethylene tube. Two heat seals are then made across the tube between the cartridge and the item, thus forming a sealed package that may be cut off the main tube and taken away. In this way, single items may be removed from the isolator at intervals, without the need for sterilisation or decontamination via a lockchamber or RTP. This can carry on until the layflat tubing is exhausted and another batch must be loaded onto the cartridge, whilst the door to the isolator is closed, as described by the manufacturer.

If the application is toxic, then new tubing can simply be pulled onto the cartridge and the cartridge fitted back onto the port door on the isolator. If the application is sterile, then the tubing and the interior of the cartridge must, of course, be sterile. This may be achieved by gamma irradiation of

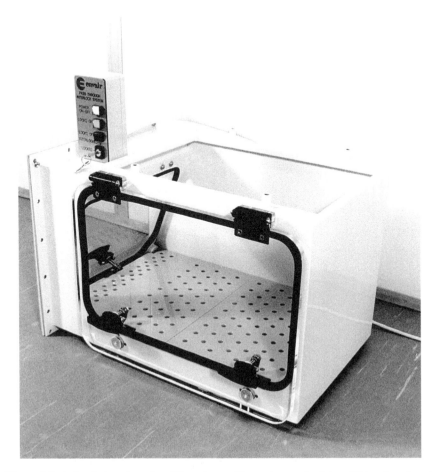

Figure 3.5 A Lockchamber Awaiting Assembly onto an Isolator. This chamber is incomplete and has yet to be fitted with a panel HEPA filter on the roof, but it will be of the D class. It features an electromagnetic interlock system, which prevents the opening of both doors at the same time. (Courtesy of Envair Ltd.)

the bulk tubing, either together with the cartridge or separately. Irradiated tubing may be fitted to a nonsterile cartridge fitted to the isolator, after which the isolator door may be opened and the exposed tube and cartridge interior sterilised by gassing.

Several versions of this port are available, one being based on an RTP container body, which can be undocked from an RTP on the isolator and replaced with a fresh unit, presterilised if required. Another version resembles a lockchamber in that the cartridge is housed in a sealed enclosure with doors into the isolator and out to the exterior. This design overcomes the problems of isolator overpressure inflating the polyethylene tubing and pulling it off the cartridge or underpressure drawing it into the isolator. Figure 3.7 is an example of a product passout port. Note the cartridge and a pack

Figure 3.6 A Lockchamber for a Rigid Isolator. A lockchamber is mounted on a rigid wall isolator and fitted with a pull-out drawer to make loading and unloading of the chamber from within the isolator more ergonomic. (Courtesy of Astec Microflow.)

of polyethylene layflat tubing, ready to be fitted. Figure 3.8 shows a product passout port ready for use.

Waste ports

Somewhat akin to product passout ports are the various designs of waste removal ports, based on bagging techniques. One example has the waste bag sealed to a ring that carries large O-ring seals, allowing it to pass through

Figure 3.7 A Product Passout Port Fitted onto a Flexible Film Glove Isolator. The white plastic cartridge is loaded with the pack of polyethylene layflat tubing in the foreground, then attached inside the chamber. (Courtesy of Astec Microflow.)

Figure 3.8 A Product Passout Port in Use. This in-use product passout port shows the layflat polyethylene tubing in place ready to accept items from the isolator. In this particular application, the user ties off the tubing rather than using a heat sealer. (Courtesy of Astec Microflow.)

an outer ring mounted in the isolator floor. When the bag is full, a new bag and ring are used to displace the old assembly, out of the isolator, the isolator seal being maintained as the new assembly takes over from the old one. This design is primarily for use in sterile applications (e.g., hospital pharmacy units). Other types use double-bagging methods and bag-over-bag techniques, though these designs are perhaps obsolescent today. Figure 3.9 illustrates a waste port with telescopic bag-change capability.

Figure 3.9 A Waste Port in the Base Tray of a Flexible Film Isolator under Assembly. In this type of port, the waste bag is changed with a telescopic ring system. The picture shows the lid has been removed, and a new bag ring is ready to be pressed down into place, dispensing the old bag and ring out of the isolator.

Rapid transfer ports

The RTP or Double Door Transfer Port is one of the most sophisticated transfer devices available to the isolation technology designer. It can be used in a wide range of transfer applications, from liquid products to solid wastes, in sterile and in toxic fields, but, as discussed in Chapter 1, the RTP is not the raison d'être of isolation technology that some would have us believe. It is an excellent device and irreplaceable in many contexts, but it is only one of many devices that can be applied to transfer into and out of isolators. Figure 3.10 shows the author operating an RTP.

The RTP was developed for the nuclear industry around 30 years ago, mainly by the French company La Calhène SA. Much effort was given over to optimising the design of seal and the tolerances of the mechanical parts, and tooling was produced to injection mould some of the components. It can probably be said that, whilst a number of similar ports have been made by other manufacturers, at the time of writing, the La Calhène SA port is one of the finest available.

Figure 3.10 The Author Operating an RTP. The alpha flange or female port is mounted on a flexible film glove isolator. The beta flange or male section is fitted to a moulded plastic box to form the rapid transfer container. The container is docked onto the port with a bayonet action. (Courtesy of Glaxo Manufacturing Services Ltd.)

Mode of operation

The RTP consists of two main assemblies, the container section and the port section. These are sometimes referred to in Europe as the male section and the female section, whilst the Americans, who are chary of such allusion, refer to them as the alpha and beta sections.

The terminology in this book is as follows:

- Container section (equivalent to male or beta section), consisting of a container flange, a container lid, and a container seal.
- Port section (equivalent to female or alpha section), consisting of a port flange, a port door, a port seal, and some form of closing mechanism.

Normally, the port will be fitted to an isolator wall and the container will be a mobile unit, though this may not always be the case. The way in which the RTP functions is best understood from Figure 3.11 to Figure 3.13.

The container flange carries a lid that is made gas-tight by a specially shaped seal. The port has a door, also made gas-tight by an equivalent seal. The container is presented to the port and docked on with a bayonet action, rotating clockwise. This rotation simultaneously carries out three functions:

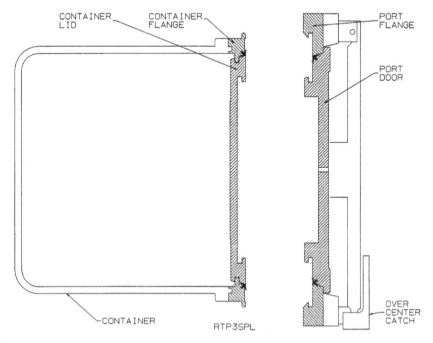

Figure 3.11 A Cross-Sectional Drawing of an RTP and Container. This figure shows the container and the port separate, prior to docking the two together.

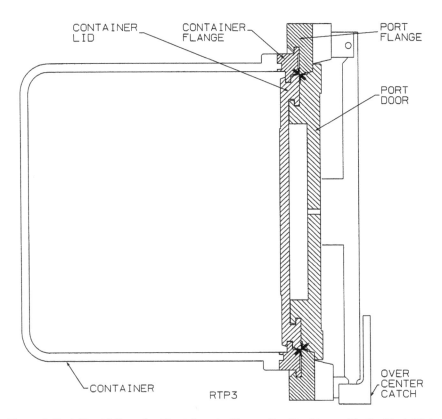

Figure 3.12 A Rapid Transfer Container in Place after Docking with the Port. Note how the arrowhead seals on the container and the port meet point to point.

the container flange is locked onto the port flange, the container lid is released from the container flange, and the container lid is locked onto the port door. Note in Figure 3.12 how the two arrowhead seals meet up point to point, and thus bring, to as near zero as possible, the common area that could transmit contamination. Note also that not all RTP designs utilize rotation for docking the container to the port. The IDC version of the RTP uses a system of radial locking pins to both hold the container flange to the port and to release the lid from the container.

From within the isolator, the port door is opened, and with it comes the container lid, their two exterior surfaces now sealed together by the door seal. All surfaces now exposed within the isolator and the container are common and may be sterile or toxic, depending on the application; thus, transfer can be made between the isolator and the container without breaking containment. Once transfer is complete, the port door is closed from within the isolator and the container may be then undocked from the outside.

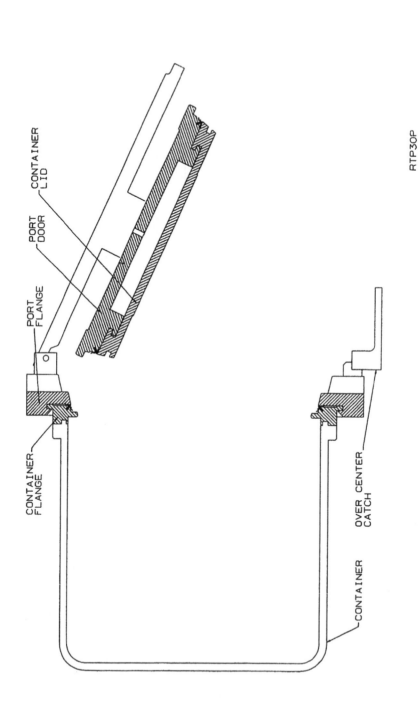

CONTAINER LID

PORT DOOR

PORT FLANGE

CONTAINER FLANGE

CONTAINER

OVER CENTER CATCH

RTP3OP

Figure 3.13 An Open Port Door. In this drawing, the port door has been opened, bringing with it the container lid, their contaminated surfaces being sealed gas-tight. Transfer can now take place without further sterilisation or decontamination.

Operational considerations

Occluded surfaces

In sterile applications, the container and the isolator must be sterilised, and, indeed, the periphery of the port door and the container lid must also be treated, as these become exposed when the port is opened. This may be addressed by sterilisation of the isolator with the container docked and the door open, or by manual treatment of these areas with a wet chemical agent. In toxic applications, the peripheral surfaces are not exposed to the exterior directly, but even so, they should be decontaminated on a regular basis.

"Ring of concern"

The efficiency of the RTP is dependent on the accurate presentation of the two arrowhead seals, nose to nose. Naturally, the seals cannot come to a true point, and the whole port is subject to both engineering tolerances and to wear in use. Thus, there is a ring at the nose of the seals in which contamination may potentially be exchanged. Even in the best-engineered port, this ring might be 0.1 mm across, and in a port of, say, 350-mm diameter, this represents a total area of some 55 mm^2, on which a considerable number of microorganisms might reside.

There are two practical ways to tackle this potential problem: (1) The bioburden (or contamination burden in a toxic application) in the vicinity of the port and container face can be minimised. (2) The seal noses may be treated directly by hand with a decontaminating agent.

Various technical solutions to the problem have also been tried, including the dry heat-treatment of the ring after docking but prior to opening the port, stream treatment, and treatment by intense visible light pulse. Whilst the devices have produced good results, they are quite large, tend to be costly, and take time to operate. The problem is to some extent mollified by nonrotative RTPs, such as the IDC version, since any contamination present is not wiped around the periphery of the seal.

Interlocking

In the plain RTP, it is possible to open the port door with no container in place, resulting in a massive breach of containment. The best RTP will have a mechanical interlock that prevents opening the port door if a container, with a lid, is not present and also prevents removal of the container if the port door is not closed. Such devices are effective, but are relatively expensive and introduce further mechanical complication. Figure 3.14 shows a La Calhène SA RTP from within the isolator. This has a neat mechanical interlock.

Figure 3.14 An RTP Viewed from Inside an Isolator. This shows an RTP viewed from within the isolator, with the port door open to show the interior of the container, which has been docked on. This port is fitted with a very neat mechanical interlock, which prevents the operator from opening the port if a container is not present. (Courtesy of La Calhène SA.)

Engineering

The RTP is a precision device and, if abused, will fail in various ways. For example, if the port, particularly if large, is bolted up against a surface that is not flat, it will distort and the container flange will not enter. As another example, the container lid is designed to pass through the port flange with minimum clearance. If the lid is dropped on its edge, a burr will be produced that can stop the lid from passing though the port.

RTP containers

The container body used with the RTP system may be of almost any shape, size, or material to suit the operation. The most common types are described below.

Simple

Simple containers are normally plastic drums or boxes of various sizes that are used to shuttle between isolators. They may be sterilised prior to use, either by docking onto an isolator that is then gassed with the RTP door open, or by placing the open container inside a suitable isolator that is then gassed. The lid is fitted back on the container before it is removed from the isolator. If the application is toxic, then the container may be decontaminated manually whilst docked to the isolator.

Gassable

Much the same as the simple types, gassable containers are fitted with small HEPA filters and connectors so that they can be gas sterilised using one of the various types of gas generators available (see Chapter 5). A typical application is for large waste containers that may be docked onto a port mounted in the floor of an isolator. Once full, the waste drum is undocked and a fresh sterile unit returned to the port.

Autoclavable

In the simple form, autoclavable containers are made from materials resistant to autoclave conditions, 134°C for 30 minutes, and are fitted with a hydrophobic membrane filter. The complete container, with the lid fitted and filled with materials, can be autoclave sterilised. The filter allows for the escape of air and the entry of steam on a typical porous load cycle. Once cool, it can be docked onto an isolator and the sterile contents transferred immediately. A more sophisticated version actually allows the lid to be lifted from the container and the entire assembly is then steamed through, so that the container seal is effectively sterilised. A typical application for autoclavable containers is the transfer of stoppers to an aseptic filling line. The body of the container may be aluminium if the process allows, it may be stainless steel (although weight can become a serious consideration here), or it may be in the form of a Tyvek™, or similar, bag. Figures 3.15 and 3.16 show an autoclavable container.

Partially disposable

In the disposable version, the container body is in the form of a bag, which is sealed to a male container flange and lid assembly. The bag may be filled with items such as vial stoppers, and then the whole assembly is autoclaved before docking onto, for instance, an isolated vial filling machine. In one version the male flange assembly is recycled after transfer; in another version, the complete assembly is used once and discarded. Recycling is expensive in terms of cleaning and validation, whilst single use is desirable for Good Manufacturing Practice (GMP) reasons but costly in financial terms. This type of RTP transfer container, known as the Beta Bag™ system, has nonetheless found wide application in pharmaceutical production.

Figure 3.15 A Stainless Steel Autoclavable Rapid Transfer Container. Note the handles used to dock and rotate the container. (Courtesy of La Calhène SA.)

Figure 3.16 The Autoclavable Rapid Transfer Container from the Back. This shows the hydrophobic filter fitted to allow air to escape from the container and steam to enter during the autoclave cycle. (Courtesy of La Calhène SA.)

Totally disposable

A more recent development has been the totally disposable RTP container and port system produced by the French company IDC. The container can only be used for one single docking cycle and is entirely disposable. This then removes the need for cleaning, decontamination, sterilisation, and the validation work associated with these activities. The containers are generally specially manufactured to contain particular process items, the bodies being fabricated from a variety of flexible plastics. The containers are quite small and are produced in relatively large numbers; thus, the costs are kept at very competitive levels. The IDC port has found early application in French hospitals but is making strong inroads into the pharmaceutical industry: at the time of writing, at least one major UK pharmaceutical company has committed itself to extensive use of the port. The port is shown in Figures 3.16a and 3.16b.

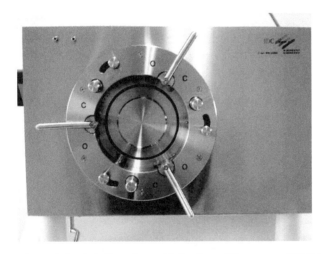

Figure 3.16a The IDC BioSafe Female or Alpha Port. (Courtesy of IDC.)

Figure 3.16b An IDC BioSafe Container with Male or Beta Flange. This container is designed to hold presterilised stoppers. (Courtesy of IDC.)

Isolator

In some cases, the container may be a complete mobile isolator, used perhaps to shuttle materials between one cleanroom and another. Such an isolator may have a battery backup (e.g., UPS) to maintain conditions during the transfer and during any storage period before transfer. A flexible plastic sleeve is needed to connect the container section to the isolator, to allow for the rotation of docking, and to give flexibility to line up the two parts of the RTP system. Figure 3.17 shows a container flange (beta flange) fitted to a transfer isolator.

Powder transfer

The problem of transferring a powder from one bulk container to another, either maintaining sterility or containing toxic hazard, is common in pharmaceuticals. This can be tackled by use of an intermediate box fitted with RTPs at the top and bottom. This can then be used to introduce a transfer pipe between the two vessels, either by hand or by automatic mechanism. When transfer is complete, the RTP doors are closed and the upper vessel can be removed directly. Figure 3.18 shows such a powder transfer device.

How good is the RTP?

Determining the quality of an RTP is a question that any prospective user of an RTP will naturally ask, and there are two aspects to the query. First is the simple question of seal, which is easy to quantify. The lid should seal

Figure 3.17 A Beta or Male Container Flange. The beta or male container flange is connected to a transfer isolator via a flexible film sleeve. The sleeve allows the operator to lift and rotate the flange to dock onto a corresponding alpha flange or female port on another isolator.

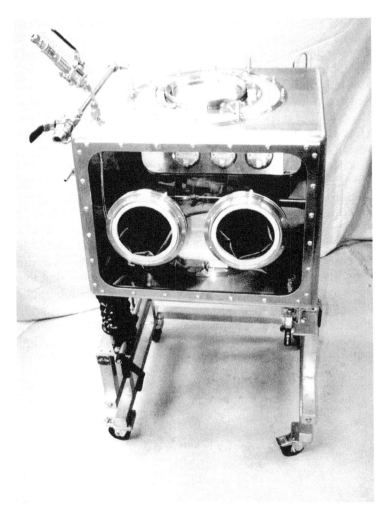

Figure 3.18 Two RTPs for a Bulk Powder. This device uses two RTPs to connect a bulk powder container with a vessel without break of containment. The gloves are used to make the connection within the chamber. (Courtesy of La Calhène SA.)

airtight to the container, the door to the port, the container to the port, and the lid to the door. These can be checked by pressure-drop tests, but the real concern is with the potential for cross-contamination during transfer. The device is not an absolute barrier to cross-contamination because of engineering limitations; it is an attenuating device. The user thus needs to determine how good the attenuating action is likely to be — a parameter that is not easy to measure directly. Most of the evidence relates to the La Calhène SA-manufactured ports, which were developed 30 years ago and have been in widespread use since then. The indications are that the port is a very good attenuating device indeed and is probably as good, in practice, as the supposedly absolute transfer devices, such as gassable lockchambers.

The initial tests of the La Calhène SA port were in the nuclear industry and involved the transfer of plutonium oxide powder. There was some criticism of the rotational action, which tends to rub material onto the surface of the seals, but the port has nonetheless become widely used in nuclear gloveboxes in Europe and in the U. S. Combined with appropriate shielding, depending on the level of activity, the port is an accepted method of transfer, particularly in waste disposal where oil drum-sized containers may be used.

The first applications in a nonnuclear field came when the port was fitted onto research animal isolators for the transfer of food and bedding and the removal of waste. It was soon found that the port could provide reliable transfer in SPF and germfree isolation units over long periods of time. Thus, the indications are that the port works very well in preventing cross-contamination during transfer.

At the time of writing, there is some direct experimental data relating to the microbiological challenge of the RTP, but not a great deal. This seems a notable omission on the part of the port manufacturers. We must rely basically on historical evidence. The conclusion is, however, that the RTP is an excellent and reliable transfer method for use in isolation applications. When asked directly how their port performs, La Calhène SA representatives respond that it is as good, in terms of attenuation, as HEPA filters used on isolators. We can conclude that this is a satisfactory situation. Figure 3.19 shows RTPs fitted onto a suite of isolators for use in a hospital pharmacy.

Figure 3.19 These Isolators Are Designed for Use in a Hospital Pharmacy for Aseptic Dispensing. RTPs and containers are used in this installation to move items from the sterile bank isolator (furthest from the camera) to various dispensing isolators. (Courtesy of Astec Microflow.)

A further type of RTP: the split butterfly port

The split butterfly port was developed by the German company Buck GmbH in conjunction with the pharmaceutical industry. In essence, it is a split butterfly valve, the butterfly plate being split horizontally. The two parts of the port are occluded by half-thickness butterfly valve plates. When the two halves of the port are docked together, the half-thickness butterfly plates seal together in much the same way as the lid-door assembly of the La Calhène SA port. The port then acts like a conventional butterfly valve and can be opened and closed by a manual lever or power actuator. The port is specifically designed for hazardous powders and chemicals in bulk, but might be adapted for sterile use in some cases. Figure 3.20 is an exploded drawing of a split butterfly port.

At the time of writing, a number of other types of RTP are under development. One of these uses direct heating of the port faces to sterilise them quickly before docking, for use in high-speed production applications. We may expect to see other versions of RTPs appear as the use of isolation technology extends.

Direct interface

Most production isolator suites will interface directly with some kind of process equipment. Interface equipment may be divided into two broad types:

1. Equipment that is serviced by the isolator, such as filling machines, blenders, and reactors.
2. Equipment that services the process carried out within the isolator suite, such as autoclaves, depyrogenating tunnels, and freeze dryers (lyophilisers).

In the first category, considerable design ingenuity may be required to make an ergonomic and practical seal between the machine and the isolator. The problems are such that it is often better not to use an existing piece of process equipment, but to buy new equipment specifically adapted by the manufacturer to interface with the isolator (Ohms 1996). This adaptation may simply be the addition of a suitable mounting flange onto which the isolator may be bolted, but it may involve fundamental changes to make the equipment more compact and to fit shaft seals, for instance, where mechanical drives must pass the isolator wall. The problem is becoming less acute, however, since the manufacturers, particularly of filling machines, have become alert to the possibilities of isolation technology and are now designing with isolation in mind. Even so, it will be important for the equipment manufacturer and the isolator designer to work closely together if machine and isolator are to unite successfully. An example of such a cooperation is described in more detail in Chapter 8. Figure 3.21 shows the complex integration of a filling machine and an isolator system.

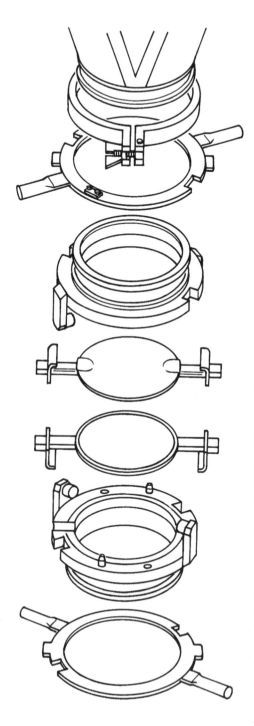

Figure 3.20 An Exploded Drawing of the Buck Port. The split butterfly valve halves are the third and fourth items from the left in the assembly.

Figure 3.21 A Complete Filling Line Isolator System. This photograph shows some of the technical complexity that may be required with such installations. (Courtesy of La Calhène SA.)

In the second category, the interface is often less intimate and thus less difficult to realise, but cooperation with the equipment manufacturer may still be useful. Problems associated with the interface to autoclaves, for example, are less to do with the sealed junction of the two units and more to do with sterilisation and decontamination. If, for instance, the autoclave has a hinged door, then the swing of this door into the isolator must be accommodated. On the other hand, if the autoclave has a sliding door, how will the door be sterilised or decontaminated so that no untreated surface enters the isolator? Again, some manufacturers are beginning to design for isolation, and one autoclave manufacturer uses the autoclave door as a drawbridge over which to convey items into the isolator. Figure 3.22 is of an isolator used to unload an autoclave.

The depyrogenating oven or tunnel is a device that may well be interfaced with a filling line isolator, and the tunnel version presents a particular problem for sterilisation. The depyrogenating tunnel is an open-ended device designed to deliver vials into a cleanroom. A door will be needed at the isolator wall to close off the tunnel whilst the isolator is gas sterilised. The volume between the door and the hot section of the tunnel is thus neither gassed nor heated and must be sterilised by some other method. One possible solution is to treat this area by manual spray-down via fitted sleeves

Figure 3.22 This Special Half-Suit Isolator Is Used to Unload an Autoclave. (Courtesy of La Calhène SA.)

and gloves. Perhaps future depyrogenating tunnels will be built with a facility to close off at the input end during the gas sterilisation phase, so that the tunnel is treated at the same time as the isolator and no door is required at the isolator wall.

Dynamic mousehole

If a continuous operation is placed inside an isolator system, then there will be a requirement for the continuous input of materials and the output of product, both of which must take place without compromising the integrity of the isolator. Direct interface may take care of, for example, the input of vials for a sterile filling process, but the output of the filled, stoppered, and capped vials presents an interesting challenge. One method might be to accumulate batches in reservoir isolators that are undocked, emptied, resterilised, and then docked once more onto the filling isolator, using perhaps an RTP. This process would be labour-intensive and would require much space, so some form of continuous output device is the only practical solution.

Clearly, the first stage is to design an output aperture of minimum size to pass the vials, the so-called mousehole. Any conveyor belt system carrying the vials cannot, of course, cross the isolator wall, since the return belt would be contaminated by exposure to the outside atmosphere. Thus, a deadplate will usually be fitted here, the vials being pushed across the plate and onto the receiving conveyor simply by the pressure of vials from the feed conveyor. Smaller vials are very unstable and can create problems with this kind of exchange, so the design may place the feed and receiving conveyors side by side for a short length, the vials being translated sideways across the gap between the two. The wall of the isolator is then established along the line between the two conveyors.

Having devised a method to move the vials across the isolator wall, we need to give careful consideration to contamination control through this port. The passout hole will be of minimum size and the positive pressure of the isolator, in sterile mode, will lose air continuously to the outside. Even so, there is potential for atmospheric exchange in this open type of isolator, as defined by Carmen Wagner (1995). This potential for exchange through apertures in isolators might take various forms, such as micro-turbulence or induction leakage, the latter resulting from the dynamic flow situation.

Induction leakage is not only of concern at the mousehole, which is a known leak site, but is also of concern at the site of unknown leaks. Taking the example of a positive-pressure sterile isolator, the logic runs as follows:

1. Air movement within the isolator possesses a dynamic pressure according to the formula: d = density in kg/m and u = velocity in m/sec.
2. If this dynamic pressure exceeds isolator pressure, then air will be induced into the isolator from the outside, through any aperture, be it a mousehole or an unintended leak.

This is a physical law, but it can be noted from the formula that the internal air velocities need to be quite high to produce induction. For instance, an air velocity of 5 m/sec gives a dynamic pressure of 15 Pa, while the standard unidirectional downflow velocity of 0.45 m/sec gives a dynamic pressure of only 0.12 Pa. On the other hand, it can be argued that the apparently safe isolator pressure of 50 Pa can be reduced momentarily to near zero by the rapid withdrawal of a sleeve. Under these conditions, the internal air velocities might be sufficient to give a dynamic pressure exceeding the isolator pressure, in which case outside air will be drawn in. If the sleeves were to be regularly withdrawn rapidly by overzealous or hard-pressured operators, then regular induction of air might take place to the obvious detriment of the isolator environment.

The converse is true for negative-pressure containment isolators where, if the airflow outside the isolator is rapid, the outward leakage of contaminated air might be induced.

To reduce the possibility that particles might enter the isolator by induction through the vial aperture, the area is surrounded by a small enclosure, itself supplied with HEPA-filtered air at a pressure midway between the isolator overpressure and the outside atmosphere. Thus, we have a dynamic form of output port, hence the term *dynamic mousehole*, which, although strictly speaking, is directly open to the outside atmosphere, constitutes a very effective barrier to the entry of microorganisms. However, it has been pointed out that the dynamic mousehole does not present a barrier to macroorganisms such as mites, and so some consideration should be given to such issues in the design and running of the room that houses an isolator that is so equipped.

Another aspect that will require consideration is the closure of the mousehole during cleaning and sterilisation, particularly if CIP/SIP with gas-phase sterilisation is employed.

Figure 3.23 shows the connecting flange used to link a flexible film half-suit isolator with a medium-sized freeze dryer. Figure 3.24 shows a flexible film glove isolator linked to a laboratory-scale freeze dryer. Figure 3.25 shows a miniature freeze dryer connected through the base tray of a small flexible film isolator. Figure 3.26 shows the direct connection of two biomedical isolators to a Class 2 biological safety cabinet.

Pipe and hose connection

Another type of transfer is the delivery of liquids directly to an isolator, a fairly common requirement in both toxic containment and sterile processing. Where the volume of liquid is small, for instance, less than 5 L, perhaps the best plan is to take the vessel directly into the isolator via the lockchamber of an RTP. Larger volumes of liquid will require a different approach. If the application is one of toxic containment, then a simple method is to fit a tube through the isolator wall with hose barbs inside and outside, and a suitable cutoff valve, probably inside the isolator. Silicone rubber or similar suitable material tubing is used to connect onto the outside of the isolator, with a further length inside to reach, for instance, a filling machine. When the process is complete, the whole line is flushed with a solvent or deactivating solution, after which the valve is closed. The inner piece of hose is bagged for disposal inside the isolator and the outer piece bagged outside.

If, however, the process must be sterile, then there is an interesting problem in making connection with the isolator from the outside. One of the simplest methods may be "flaming on," in which the end of the hose from the product vessel, probably previously sterilised by autoclaving, is pushed onto a hose barb on the outside of the isolator whilst covered by an alcohol flame. A slightly more sophisticated method is to push the end of the previously plugged and autoclaved hose through a compression gland in the wall of the isolator. The isolator is then gas sterilised, the porous plug is removed, and the hose is connected as required inside the isolator.

A well-engineered solution to this problem has been devised by La Calhène SA, using RTP technology (see Figure 3.27). In this method, an autoclavable RTP container is fitted with a double hose barb in the base. The outer barb carries a length of silicone rubber hose connected to the product vessel; the inner barb carries a further coiled length of tubing, and perhaps a line filter as well, all within the RTP container. This whole assembly of

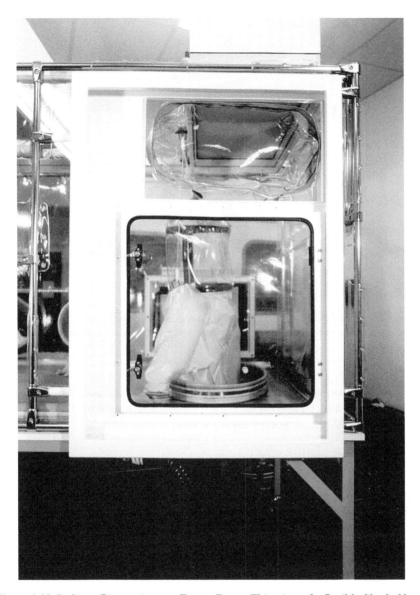

Figure 3.23 Isolator Connection to a Freeze Dryer. This view of a flexible film half-suit isolator shows the flange that will be used to connect it to a medium-sized freeze dryer. (Courtesy of Envair Ltd.)

container, tubing, and vessel is then autoclaved or steamed through, depending on the design. Now the container can be docked onto a matching port on the isolator, the door opened, and the sterile tubing pulled out to connect in the isolator as required. When the product vessel is empty, the tubing is disconnected and returned to the RTP container from within the isolator, the container is undocked, and another can immediately take its place to

Figure 3.24 A Pilot-Scale Installation of a Glove Isolator and a Small Freeze Dryer. Note how the door of the freeze dryer opens out into the isolator. (Courtesy of Astec Microflow.)

continue the process. If transferring liquid between two isolators, then two RTP containers may be used at each end of a suitable length of silicone rubber hose. Figure 3.27 shows such a liquid connection system.

Services

Many applications will require services within the isolator, especially mains electrical power supply. Suitable cleanroom electrical sockets, such as the Clipsal™ type, can be fitted, but all sockets are complex in terms of the cavities and crevices they contain, and so have implications for both sterile environments and for toxic containment. Where practical, the simple compression gland to carry an electrical cable provides a more crevice-free solution for power supply.

Similarly, any other service, such as gas, vacuum, or compressed air, must be engineered to match the isolator environment. If compressed air is used in the isolator, then it should be exhausted outside the isolator or the exhaust passed through a filter. Hydraulic systems are generally to be avoided in isolators. Figure 3.28 shows electrical sockets built into the base tray of a glove isolator.

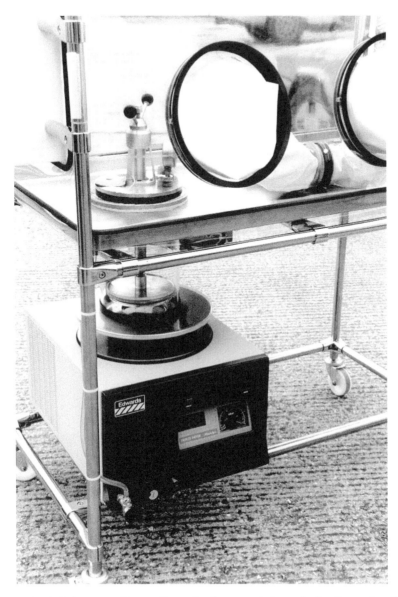

Figure 3.25 A Laboratory Freeze Dryer Is Connected through the Base of a Glove Isolator.

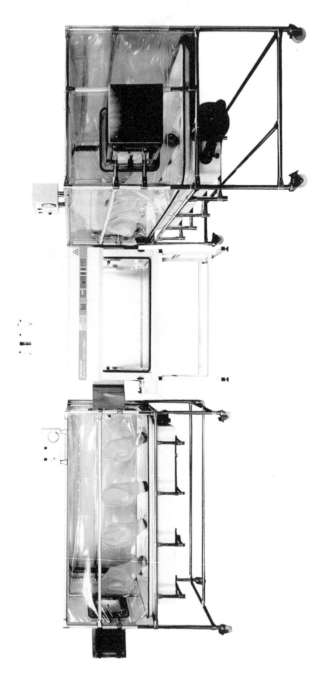

Figure 3.26 Two Biomedical Isolators Connected to a Class 2 Biological Safety Cabinet. (Courtesy of Astec Microflow.)

Figure 3.27 The Aseptic Connection of a Product Hose to an Isolator Using a Small RTP. In this case, the connection is through the base tray of the isolator. The stainless steel autoclavable RTP container is seen docked below the tray. The port in the tray is open, and the product hose bends away to the left of the picture. Note the hydrophobic filter fitted to the bottom right of the container. (Courtesy of La Calhène SA.)

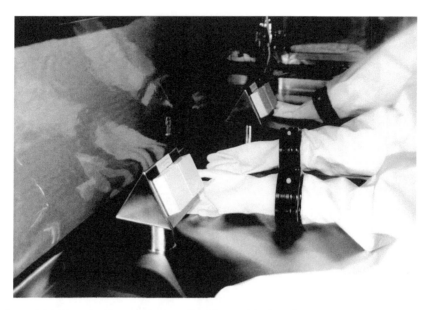

Figure 3.28 Electrical Power Sockets Fitted to a Glove Isolator. The sockets are fitted to the glove isolator in stainless steel "Christmas trees" to elevate them above any liquid spillage. (Courtesy of Astec Microflow.)

chapter four

Further design considerations

Isolator control systems

The control system fitted to an isolator may be a simple, manual type or it may be a sophisticated, programmable logic controller (PLC)- or personal computer (PC) -driven unit; in either case, the aim of the control is to set up and to maintain the environmental conditions specified for the process in hand. Generally speaking, the two major parameters that must be maintained are the isolator pressure, be it positive or negative with respect to the isolator room, and the airflow rate, be it turbulent or unidirectional. More complex systems will include controls for temperature and humidity, and some systems may be required to control more exotic parameters, such as oxygen content (Figure 4.1 through Figure 4.3).

One important factor to remember: if the isolator control system is run through a PLC or PC, the software developed will require validation if the isolator setup is to be used for licensed manufacture. The recognized standard for the validation of computer-controlled systems is Good Automated Manufacturing Practice (GAMP), further discussed in Chapter 8. There are specialist companies who provide this service, and it can represent a sizeable proportion of the cost of an isolator system. Potential users must be aware of this factor.

Simple, turbulent flow isolators

Simple, turbulent flow isolators will normally be no larger than a four-glove unit and will have only a single fan to drive the ventilation system. They may be either positive or negative pressure. In the positive case, the fan will be on the inlet side, and in the negative, on the exhaust side. The fan is likely to be a small centrifugal type of no more than perhaps 150 W. This type of isolator is easy to control since the system is inherently stable, as described below in "Flow." The fan may be sized to run at full speed continuously, or

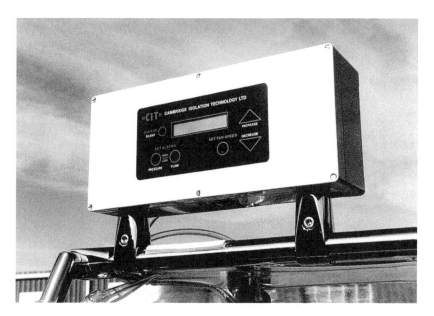

Figure 4.1 The Control System Fitted to a Flexible Film Isolator. The touch panel gives a digital display of the isolator canopy pressure and the airflow rate through the isolator. It also allows the operator to set the required canopy pressure and the alarm levels for high and low excursion of the pressure and airflow. A neat and simple system.

a damper valve may be fitted to control the airflow. Most likely, however, the fan will be fitted with a speed controller. This could be a manually adjusted Variac transformer or a thyristor (lamp-dimmer) circuit, or it may be a fully automatic electronic control system with feedback from the pressure sensor or possibly the flow sensor.

Double-fan turbulent flow isolators

It is common to fit two fans onto an isolator, one on the inlet and one on the exhaust, so that pressure and flow can be controlled by the relative adjustment of the fan speeds. This also obviates the need for very large filters; the exhaust filter of a single-fan, positive-pressure isolator is driven by the canopy pressure, and the converse is true for negative pressure. If a relatively large airflow is required, then a much smaller HEPA filter can be used if fan pressure is available to drive it.

The problem with double-fan systems is that, as mentioned in Chapter 2, they are dynamically unstable. Small changes in the speed of one fan can have a major effect on the pressure in the system as the second fan responds to the alteration. Thus, the control system governing fan speed must be precise and have stability better than 0.1 percent of the full power rating. If the control is manual, then very stable circuitry will be needed, impervious to changes in temperature, input voltage, etc. If the control is electronic, some

Figure 4.2 A Very Simple Monitoring System Fitted to a Biomedical Isolator. The beacon on the top of the enclosure provides a silent alarm if the pressure in the isolator exceeds the setting of 80 Pa on the Photohelic gauge. The smaller gauge gives a visual indication of the airflow through the isolator. (Courtesy of Astec Microflow.)

Figure 4.3 The Control and Instrumentation System Fitted to a Zone 1 Flameproof Isolator. Again, the isolator pressure and airflow are the parameters of concern while, in this case, the alarm levels are internally fixed and not operator defined. (Courtesy of Astec Microflow.)

care will be needed to develop algorithms that do not have harmonics similar to any periodicity in the complete system, including any exhaust ductwork. Probably the simplest way to tackle the problem is to fix the speed of one of the fans to give, for example, a known flow rate, and then vary the other to arrive at a set isolator pressure. Some precautions when considering the use of double-fan arrangements are given below in "Flow."

Simple, unidirectional flow isolators

Simple, unidirectional flow isolators may be positive or negative pressure; they may exhaust to the room, or the exhaust may be ducted to the atmosphere. In either case, the design problems are similar to those for turbulent flow, except that the flow rates are one or two orders of magnitude greater. Exhaust to the room makes control a little simpler, especially for the double-fan case, but the fans will dump an appreciable heat burden into the room. Ducted exhaust to the atmosphere, on the other hand, will remove a lot of air from the room that the HVAC system must replace. If there is a remote fan at the efflux of the exhaust, in order to maintain negative pressure in the exhaust ductwork, then control will be more complex.

Recirculating unidirectional isolators

Recirculating unidirectional isolators are often fitted to pharmaceutical filling lines, and so they form an increasingly important group. Whilst the flow diagram of such systems appears complex, the control problems are perhaps not as intractable. The flow rate, which will probably be governed by the requirement to have a velocity off the inlet HEPA filters of 0.45 m/sec, can be controlled by the main circulation fan or fans. The pressure in the isolator chamber may then be handled by controlling the input and output of makeup air. The makeup air may have a single- or double-fan arrangement that may be governed manually, but is more likely to be under the command of a PLC or PC. The single-fan design is probably an easier prospect since it will be dynamically more stable.

Temperature and relative humidity

Since it is the remit of isolators to provide specific atmospheres, temperature and relative humidity are often defining parameters. Both can be measured fairly easily, but active control is not as simple. Refrigerant/heating systems can be fitted to the air supply for the isolator and are often fitted as a matter of course on many recirculating units. However, isolators tend to be very good heat exchangers, so they equilibrate quickly with the surrounding room environment. For this reason, it is not easy to control a temperature that is more than one or two degrees different from the room. If a larger differential is required, then the isolator will need insulation — not easy where windows and sleeves are in use. Relative humidity may be nominally easier to control, but since it is directly related to temperature, the same considerations apply. Given the resources, most requirements can be achieved, but potential users should be aware of these limitations.

Isolator instrumentation

As with control, instrumentation may be very simple and consist of straightforward analogue gauges, or it may be a sophisticated array of parameters derived from transducers and linked to process equipment, building management systems, alarms, and the like. The degree of sophistication will depend on the level of monitoring dictated by the process to be carried out in the isolator. As with much instrumentation, generally, it is not so much the absolute values that are of concern, but changes from the established norm that must be monitored. Even so, the instrumentation will require calibration to known standards, particularly where licensed manufacture is concerned. The supplier should be prepared to provide this service — to National Measurement Accreditation Service (NAMAS) standards in the UK, or to any other local standard as required. Most pharmaceutical users will list equipment instruments and classify them as noncritical or critical. They will then establish the tolerance and frequency of calibration required, giving each transducer and meter a unique plant number.

The two basic performance parameters of any isolator are the pressure inside the isolator and the airflow rate through it.

Pressure

In sterile applications, it will be necessary to ensure that positive pressure is maintained at all times, within certain limits, to ensure that ingress of contamination cannot take place. Conversely, in toxic containments, the operator will need to know that negative pressure has been maintained. For these reasons, the isolator pressure meter is normally designated a critical instrument. The ubiquitous Magnehelic gauge is often used to indicate isolator pressure; this is a simple but sensitive mechanical pressure-measuring device that has a needle-and-dial analogue display. Calibration may be in inches of water gauge or Pascals, as preferred. The Photohelic® version has relays fitted, and these can be used to provide a signal when the pressure has reached certain high or low values. These are set by the operator using the control knobs on the front of the instrument. If an analogue electrical signal is required, then a pressure transducer may be fitted to the isolator, the signal being fed to the control system for digital display, alarm monitoring, or process control. Some very accurate and extremely sensitive pressure transducers are now available on the market. In either case, a line filter should be fitted to the tube, connecting the isolator with the transducer, close to the isolator wall. Apart from separating the dead end of the tube from the isolator, it will provide a degree of mechanical damping against the transient pressure changes caused when the operator enters or leaves the sleeves or half-suit.

Flow

In the recent past, isolator manufacturers have tended not to fit instrumentation to demonstrate the airflow rate through their isolators. This has been a notable omission, since the airflow through an isolator is the single most influential parameter defining the conditions within the isolator. Airflow rate is likely to have a much greater effect on air quality than the isolator pressure, but unreasonable emphasis has often been placed on pressure measurement at the expense of flow measurement. Perhaps one reason for this is that airflow rate is not easy to measure at low pressure and flow rates. The most precise method is to apply a suitably sized orifice plate and then measure the pressure drop across the plate using a pressure transducer. The airflow rate or, more precisely, the mass flow rate, can be related to the pressure drop by mathematical formulae; specialist companies can provide the equipment and the calibration data. One problem with this arrangement is that it introduces a large pressure drop in the ventilation system, requiring fan power to overcome this. Mechanical devices, such as vane anemometers, might be used, but such instruments are not designed for continuous use and also introduce resistance to flow. Probably the best method is to use a version of the hot-wire anemometer, usually using a small thermistor as the heated element. This can give an electronic signal that may be calibrated by using a standard orifice plate during setup. This flow transducer can be used to sense velocity off the face of a HEPA filter for unidirectional flow or to measure the overall flow in the ventilation system. Pitot tubes linked to pressure transducers may also be used to measure flow rate, but they tend to require relatively high duct velocities to give a satisfactory signal-to-noise ratio. As with pressure, flow rate information may be linked to alarm systems, building management, or process control. Recirculating isolators will very likely feature instrumentation of the main flow as well as the makeup flow. The isolator airflow rate meter is normally designated a critical instrument.

HEPA filter pressure drop

As with biological safety cabinets and laminar flow benches, it is common to measure the pressure drop across HEPA filters in isolators. The normal convention is that once the filter reaches a pressure drop twice that of its original clean state, it should be changed. The same instruments may be used here as on the isolator pressure measurement, either mechanical or electronic. The need for HEPA pressure-drop measurement (Figure 4.4) is not quite so critical in isolators, provided that the isolator pressure and flow are monitored. A blocked HEPA filter, either inlet or exhaust, will show up as changes to pressure and flow rate that will indicate a problem needing investigation; thus, the filter pressure-drop meters do not need to be designated as critical instruments.

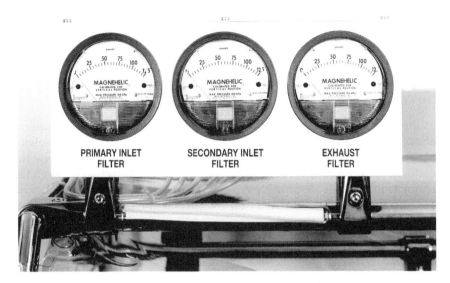

Figure 4.4 Gauges Fitted across HEPA Filters on a Flexible Film Isolator. The gauges show their condition. The client chose to fit double inlet filters in this application for added sterile product security. (Courtesy of Astec Microflow.)

Temperature and relative humidity

Temperature and relative humidity are commonly monitored in isolators, since the various cleanroom atmosphere standards (e.g., BS 5295, ISO BS EN 14644) indicate limits on these parameters. A variety of temperature and relative humidity probes are available, and the only major consideration is the siting of the sensor to give a representative reading. Solid-state relative humidity probes may be sensitive to sterilising agents, and the optimum place for these is probably in the exhaust duct, beyond the sterilising gas output connection. The readings may be simply monitored, or they may be fed back to control the isolator HVAC system, through a PLC or PC. Temperature and humidity meters may be designated as critical instruments if temperature and humidity have a specific bearing on the process being carried out in the isolator.

Particle counting

Some isolator specifications will call for in-built monitoring of other parameters, and particle counting is a popular choice. Suitable single-point or multipoint manifolds are available, and the question of optimum siting arises again, together with protection during gas sterilisation. It should be noted, however, that such in-built particle counters are expensive, and many users will not wish to dedicate the instrument to this continuous use. An alternative is to build in the sensor head and make connection to the counter when required, though this leads to problems in making the connection without breaking containment. Clearly, a line filter cannot be used here.

Air sampling

A number of air-sampling systems designed especially for isolator application are now available and might be considered as forming part of the isolator instrumentation. These sampling systems consist of a fixed head inside the isolator, which carries a very fine filter membrane, a connection through the isolator wall, and an air pump mounted outside the isolator. Particles collected by the filter may be analysed by various methods to establish the quality of the isolator air.

Building interface

Whilst isolators do not have the same radical impact on building design that cleanrooms or containment suites will have, it is clear that consideration should be given at an early stage in any isolator project to building requirements. If the application is for licensed sterile manufacture, then it would be prudent to design for a European grade D room environment — to be discussed further in Chapter 5. If the application is considered highly toxic, then it would be good practice to design for a negative pressure within the isolator room and to put HEPA and/or carbon filters on the room exhaust air.

   If the isolator system will require an exhaust duct, then this should be as short, wide, and straight as practical. Where a terminal fan is to be fitted, then control connections back to the isolator may be required. If the isolator is to be connected to a building management system (BMS), then connections should be provided.

   Power supplies, drains, or other services may also need to be designed in at an early stage so that installation and commissioning can proceed easily. A final consideration should be access. Large, stainless steel isolators cannot be dismantled and can be very heavy, so the method of access and entry to the room must be large enough to accommodate the isolator structure and any lifting gear.

chapter five

How to draw up a design specification

At this stage, the reader will have some idea of the history of isolation, its application, and its design. This is probably a good time to start thinking in terms of how to approach particular projects, and how to draw up a specification that would form the basis of discussion with potential isolator suppliers. It is also the time to bring in the concept of the qualification documents, since these may well underpin the whole project. All isolation projects will require some form of validation before operations can start, and many will have to be validated by the FDA, the MHRA, or other local authority. Even if the project will not require officially recognised validation, it is still a good plan to develop the project along these lines. An important part of the overall validation process is the so-called qualification documentation, consisting of design qualification (DQ), installation qualification (IQ), operational qualification (OQ), and performance qualification (PQ). The IQ and OQ are discussed in Chapter 6 and the PQ in Chapter 8.

Figure 5.1 shows how the various documents fit into the overall process of validation. This is a simplified version of a more major chart describing the many components of the total project validation process, which appears in Chapter 8. The simpler version is used at this stage to introduce the concept; it is a very useful piece of information for those planning a licensed production system and should be followed closely.

User requirement specification (URS)

Whilst the qualification documents as a whole need to be considered from the outset of any project, the first documentation to be developed should be the URS — also known as the User Requirement Statement and more recently as the User Requirement Brief (URB.) The URS document, normally written by the client pharmaceutical company, describes what is required of the isolator or isolator system. It may suggest possible technical solutions and in some cases may dictate some absolute requirements, but its general

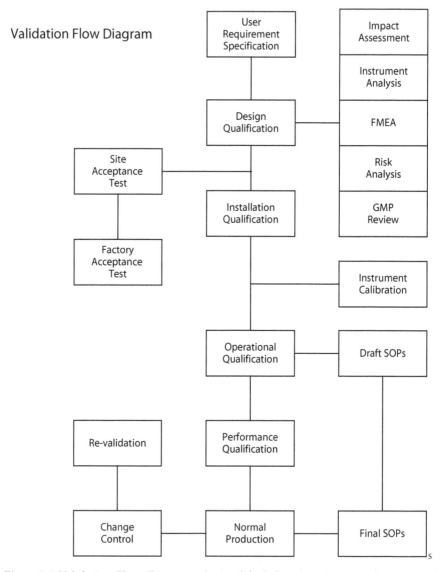

Figure 5.1 Validation Flow Diagram. A simplified drawing showing the series of documents that need to be prepared for the formal validation of an isolator, or indeed many other types of pharmaceutical installations.

purpose is to delineate how the system must perform. This document forms the basis for discussion with potential isolator supply companies. The URS may be very brief, or it may go into extensive detail, but, in either case, a clear, logically sequenced URS will speed the project to completion.

The URS forces users to think clearly about what it is they are seeking to achieve and how this might be carried out. Once this document is in place, the project can move on to produce a design with a reasonable degree of confidence that the major design requirements are at least defined, if not actually agreed upon. As with any pharmaceutical process, users have a habit of changing their minds frequently and, by setting the design parameters in the URS, the project designer can hope to reduce subsequent changes to a minimum. This chapter is largely concerned with demonstrating how to write a good URS and thus how to initiate good dialogue with potential isolator suppliers and with the appropriate regulatory authority.

Design qualification

The DQ exercise is designed to test the proposed design for conformity to the URS and, perhaps at the same time, conformity to GMP. It has been said that the validation of the system actually starts with the DQ. This is also the time at which to nominate who will be in charge of the validation process.

Standards and guidelines

One of the most significant developments for isolator technology in the last five years has been the introduction of relevant standards and guidelines that were previously lacking. Indeed, there now seems to be a profusion of studies and monographs, all of which the potential isolator user may need to reference. A comprehensive review of all of these works is outside the scope of this book, but some guiding comment is offered on each.

- BS EN ISO 14644–7, Cleanrooms and Associated Controlled Environments — Part 7: Separative Enclosures, March 2001. At the time of writing, part 7 of ISO 14644 exists as a Draft International Standard (DIS), but is likely to be promulgated with little further change. It covers aspects of contamination control other than cleanrooms and ranges from unrestricted air overspill to high-integrity inert gas gloveboxes. It is, however, entirely separate from the new standards for microbiological safety cabinets. The normative section contains nine sections, followed by seven annexes, which are designated informative. There is much useful information for isolator users in this standard and, indeed, it could provide a good template for writing an isolator URS. Normative section 5, Design and Construction, gives no values for the various parameters, but provides a comprehensive checklist for the isolator designer to follow through. Curiously, no mention is made of the need to instrument isolator airflow rate, which, as previously mentioned, must surely be a fundamental parameter of isolator operation. Subsequent sections cover Access Devices (sleeves, half-suits, and the like) and Transfer Devices. Section 9 covers testing and approval, though leak testing is discussed in Annex E. Annex B1 carries very useful information on air handling systems while Annex B2 describes gas systems, rather incongruously combining inert gas systems with gaseous decontamination systems. These two aspects are surely quite different entities and gaseous decontamination, especially using hydrogen peroxide vapour, merits a section of its own. Annex C has useful descriptions of gloves, their materials, sizes, and glove ports but describes only the glove-over-glove change method, when other types are widely used. The discussion of half-suits is very much centered around one particular manufacturer, which is not good practice in a standard. Annex E describes leak testing, but this author has reservations about the content and would suggest that readers look elsewhere for further guidance on leak testing.

- BS EN 1822:1998, High Efficiency Air Filters (HEPA and ULPA). This standard is, of course, devoted entirely to filters, but since these are fundamental components of most isolators, it is well worth studying the information. It will then be possible to give a clear specification for the supply of filters and for their subsequent *in situ* testing.

- PIC/S (Pharmaceutical Inspection Cooperation Scheme) Isolators Used for Aseptic Processing and Sterility Testing, June 2002. This document was originally produced as a guide for medicines' inspectors and written by a European committee. It has subsequently become a very effective guideline for anyone involved in the design, manufacture, testing, and operation of aseptic isolators. The structure of the document is somewhat curious in that it describes the relevant issues three times in successively greater detail; however, it contains a great deal of wholly practical information, clearly based on considerable experience in the field. Careful study of this document is recommended for those involved with isolators generally and with aseptic isolator operations in particular.

- PDA Technical Report 34. This monograph has been produced by a committee of the PDA which, though largely U.S. based, does include some European membership. Again, there is much useful information, which appears to be very much based on practical experience. The structure of the document seems a little quirky: Section 3 on Design and Construction is followed by a section named Functional Specification, which discusses airflow issues and leak testing. Next comes a section on Facility Requirements, followed by a section entitled User Requirement Specification, which deals with decontamination and environmental control. Section 7 then covers requalification. Such a structure makes it difficult for the novice to follow a logical sequence to understanding. However, the quality of the content makes the study of this document a very useful exercise.

- FDA Sterile Drug Products Produced by Aseptic Processing, draft, September 2002. At the time of writing, this FDA-produced document exists only in draft form. It may, however, form the basis of a guideline that will be vital to those seeking FDA approval of their isolator

systems. It is very much an operational document with the emphasis
on the facility, the personnel and training, the process components,
and both process and equipment validation. The coverage of isolators
is limited to an appendix, but the broad coverage of aseptic production from the FDA point of view makes this very useful background
information.

- GAMP Good Automated Manufacturing Practice, GAMP, 4 December 2001. This standard has been prepared under the auspices of the
  Society for Life Science Professionals, previously the International
  Society of Pharmaceutical Engineers (ISPE) and is aimed at the validation of computerised production systems. It provides flowcharts
  and examples of documentation that give comprehensive coverage
  of the validation process; indeed, GAMP could be applied to the
  validation of complete projects, not just to computer-based systems.
  GAMP is required reading where PLC- or PC-based control systems
  are applied to the isolator.

- PDA Biological Indicator (BI) Working Group, Recommendations for
  the Production, Control and Use of Biological Indicators for Sporicidal Gassing of Surfaces within Separative Enclosures, draft, June
  2002. At the time of writing, this monograph exists as a first draft
  and may well see some revision, but for those planning the use of
  gas sanitisation, particularly with hydrogen peroxide vapour, it gives
  very useful background information. The formation of the BI Working Group was prompted by unexplained failures of previously validated gassing cycles in production situations. This led to more detailed study of the BIs used and the eventual conclusion that the
  production and handling of BIs would have to be better regulated.
  The monograph lays out guidelines for these issues and will hopefully open the way to more stable and predictable gassing cycles in
  the future.

- UK Pharmaceutical Isolator Group, Isolators for Pharmaceutical Applications, awaiting publication, March 2004. The UK Pharmaceutical
  Isolator Group was formed in 1992 with the intention of producing
  then much-needed guidelines for isolator users. The group sought
  representation from all sides, but its main membership has tended
  to come from hospital pharmacy users and manufacturers of isolators. In addition, the MCA (now the MHRA) and more latterly the
  Health and Safety Executive (HSE) have maintained a watching brief,
  and several industrial users of isolators have attended meetings. The
  group produced a set of guidelines in 1994, and the small yellow
  booklet quickly became known as the *Yellow Guide*. It found wide
  application in the UK but has become dated over the years, and so
  a new and much more comprehensive guide has been produced by
  the group in 2003 (Midcalf, Lee, Neiger, and Coles, Pharmaceutical
  Press, 2003). This new booklet draws on the wide practical experience
  of the group members and gives comprehensive information on just

about every aspect of pharmaceutical isolators from design to validation, including issues such as relevant standards, siting and clothing, and physical and biological monitoring. Useful appendices give detail on training for operators, stainless steel for isolators, HEPA filter mechanisms, and the application of carbon filters. For those handling cytotoxics, the MHRA/HSE paper on the subject is reproduced in full as an appendix to this latest version of the *Yellow Guide*. It is suggested that potential isolator users should digest this guide as a practical and readable introduction to the subject before moving on to study the legally enforceable standards.

Workspace ergonomics and handling

The first aspect to be considered in any isolation technology project must be the nature of the work — what is the process that is to be carried out within the isolator environment? Indeed, can the work be carried out in an isolator at all? The broad answer to this very broad question is probably that virtually any process could be isolated, but that the exercise may not be cost effective. It may be necessary to go some way toward designing the isolator system before the point can be resolved.

Having decided that isolation is the correct route, for whatever blend of reasons, a whole series of interrelated factors affects the actual design, but we may start by assessing if the work should be handled in sleeves and gloves or by half-suit. Sleeves are always a first choice, but more work may be done from a single half-suit than from a number of pairs of sleeves, in some cases. As mentioned in Chapter 2, the reach limit for sleeves is about 500 mm and the practical weightlifting limit is only about 5 kg. The left-to-right reach of a single pair of sleeves is around 1200 mm and the accessible height is about 750 mm, given shoulder rings with a diameter of 300 mm or so, and a base-tray height of 900 mm above the floor. Sleeves should be set with the centre line approximately 1100 mm from the floor, and pairs of sleeves should be centred about 450 mm apart. If the isolator is fitted with sleeves along both the front and the back face, then the front-to-back width can be increased to around 1000 mm. All of these dimensions can, of course, be altered to suit particular operators, but they represent a good starting point for basic design.

Using a half-suit increases the operator lifting capacity to about 15 kg and extends the reach to a radius of around 1000 mm. These suits are conventionally mounted on oval flanges measuring 800×500 mm, and so the radius of operation springs from the edge of this flange. Again, the base tray needs to be about 900 mm from the floor, but shorter operators may need a platform to stand on when working, to gain maximum reach.

We now need to look at the process and, most importantly, the equipment that will be needed to operate the process, in relation to these ergonomic parameters. This would include not only major items such as filling machines (see "Major Equipment Interface"), but also minor items such as balances or hot plates, together with any containers, instruments, tools, and spares or change parts. All of these must be accessible to operators without excessive strain, within the confines of the isolator. Excessive strain covers not only the weight of any objects, but also the reaching and turning required and the frequency with which these actions may be carried out. In particular, the ergonomic design should avoid combined heavy lifting with turning movements, which may damage the operator's back.

A good plan in the early stages of developing the isolator design may be to mark out an area of open bench space to represent the isolator, and to lay out the equipment within the area. Items can then be moved around in a simulation of the process, and the overall dimensions of the isolator space required can then be developed. Remember that shelves can be fitted in the isolator and that the process need not be carried out entirely on the floor space. Those parts of the process that require close manual dexterity should clearly be placed directly in front of a pair of gloves, or a half-suit, and at about elbow height, perhaps 1100 mm from the room floor. Actions such as the occasional operation of a valve can be placed further away from the operator, near the maximum radius of reach.

If the manual part of the process is relatively lengthy and intricate, then consideration should be given to seated operation. This makes for operator comfort, but tends to reduce the radius of working, so that good use may have to be made of the space that is available. The whole issue of ergonomics and operator comfort raises another point worth consideration at an early stage in the development: Bring in the actual personnel who will work in the isolator. First, they may have firsthand knowledge of the process that may affect the ergonomics — it is not unknown for operators to modify the work for reasons of practicality, unbeknownst to the designers of the process. Second, it may be possible to tailor the isolator to the specific workers, be they tall or short, and thus optimise their comfort and the efficiency of the isolator operation. Third, it is probably good management policy to introduce the workers to what may be very unconventional technology at the design stage, rather than presenting them with perhaps a half-suit in which to work but no prior knowledge. Isolation technology is, of itself, in rather intimate contact with its workers, and due consideration should be given to this aspect.

One final point for consideration at an early stage should be safety issues. If the process is one of toxic containment, then the whole design will be geared up to this. But even in essentially nontoxic work, there may be safety issues. This may be as simple as the provision of safety barriers around raised work platforms, but it may include machinery guarding or interlocks to prevent manual intervention in moving equipment.

Rate of work

The rate at which any particular process is to be carried out will have a major impact on the overall design. It may govern the number of operators required. If the process must run fast, then several operators may need to work together, for example, side by side in a manual or semimanual filling operation. If, on the other hand, the work is more leisurely, then a single operator might move from half-suit to gloves and then back again as a research process proceeds.

The methods by which materials enter and leave the system will be largely defined by the rate of work. Continuous production will require sophisticated design; occasional batch work may need only very simple transfer methods (see "Transfer").

The rate of work may well influence the method of construction. For instance, a research environment with laboratory conditions and a relatively relaxed work schedule can probably accept flexible film construction. By contrast, an industrial-scale process is likely to be housed in more arduous conditions with production targets to be met, and so more robust stainless steel construction would be more appropriate. Control and instrumentation may also be influenced by the rate of work, and, in sterile application, the method of sanitisation must match the work rate.

The frequency with which the isolator is used, as opposed to the work rate in use, also affects design and construction. If the isolator will only be used on a very occasional basis, it will probably require much simpler engineering than a system that is in continuous operation. All of this information needs to be incorporated in the DQ if the appropriate design is to be developed.

Major equipment interface

The requirement to interface with process equipment is a central aspect in isolator specification. In many cases, such equipment has originally been designed for cleanroom application and presents a sanitary face that would conventionally be mounted in the cleanroom wall. The isolator interface may be fairly simple — for instance, a small drying oven might be required as part of the process. This would simply be presented at the isolator wall so that the oven door opens into the isolator, and the wall of the isolator would be sealed by some method around the outside frame of the oven door. The primary issue would be ergonomic, that of operator access to load and unload the oven and to clean or decontaminate it at the required interval.

On the other hand, the interface may be complex and intimate, as with isolators fitted to large-scale filling machines. In this case, it may be necessary to involve the machine manufacturer in conjunction with the isolator designer, so that the two items can be brought together with confidence. Again, ergonomics will be of concern, but also aspects such as the possible integration of the control and instrumentation of the two pieces of equipment and, in sterile applications, how sterility will be achieved and validated.

As an example of the latter consideration, freeze dryers (lyophilisers) are commonly connected directly to isolators. The freeze dryer may be steam sterilised while the isolator is separately gas-phase sterilised, but what of the periphery of the freeze dryer door that is closed during both sterilisation stages? This area will be occluded during both steaming and gassing and so needs to be addressed by other means, such as manual treatment.

Positive pressure or negative pressure: the cytotoxic dilemma

Having gone some way to defining the shape and size of the isolator, and its connection to any major process equipment, the next factor for consideration is the working pressure. As mentioned in Chapter 2, the conventional arrangement is for sterile processes to be run under positive pressure and toxic processes to be run under negative pressure. This will normally be a satisfactory solution if the process priority can be clearly defined in this way; however, it is increasingly common to work with products that are toxic and also require sterility. A good example of this is the production and dispensing of the cytotoxic (or cytostatic) drugs used in chemotherapy for cancer care. The QA manager, backed up by the FDA or MCA (now MHRA) inspector, will demand that the isolator be run at positive pressure to minimise the risk of contaminating the product, whilst the safety officer, backed up by the health and safety executive (HSE), will demand that it be run negative to reduce risk to the operators. This issue has been addressed by the MHRA and the HSE in the UK, working in cooperation to examine the use of both positive- and negative-pressure isolators for cytotoxics. The result has been a paper entitled Handling Cytotoxic Drugs in Isolators in NHS Pharmacies, HSE/MHRA, January 2003. Whilst aimed specifically at UK hospital use, the paper provides broad guidance for any cytotoxic application of isolators. Put very briefly, the paper concludes that the operating pressure of an isolator is not a major factor in defining the quality of the products or the safety of the operators. Other issues have a much more significant effect. Isolator users may operate cytotoxic isolators at either positive or negative pressure, but clear justification must be provided to support the eventual choice.

There are isolator designs that enclose a region of positive-pressure unidirectional downflow with areas of negative-pressure turbulent flow, thus addressing the pressure dilemma completely. In practice, however, these are complex, and most operators choose a specific pressure regime. The choice will be influenced by the nature of the product and the degree of its toxicity, but the convention at present is that if the product is deemed toxic, then negative pressure will be used. Indeed, current convention goes even further for the dispensing of cytotoxics in specifying rigid wall containment and unidirectional downflow as well. This seems a prudent approach to the problem, but it is not apparently supported by much real data — a situation that still prevails in many areas of isolation technology because little basic research has been carried out. Instead, the technology has tended to be manufacturer led, on a project-by-project basis. Thus, manufacturers have designed and built isolator systems to meet customer specifications, based on such experience they had at the time. Whilst most of these projects have been ultimately successful, it is likely that the lack of fundamental knowledge, experience, and standards has increased costs and delayed validation considerably.

One particular project has taken an unconventional approach to the problem of cytotoxic handling, not only in regard to pressure regime but also airflow and material of construction. The Baxter Special Pharmacy Services unit at Mount Vernon Hospital, North London, provides a regional cytotoxic dispensing service producing some hundreds of preparations each week. The isolators used for this work are flexible film, positive-pressure, and turbulent flow. This unit has been running for over 15 years and has had consistently good results. Not only have the products shown excellent quality, but close monitoring of the operating staff has shown no health problems associated with potential exposure to cytotoxic compounds. The view of the original designers was that, taking a balanced view, the risk to the patients from negative-pressure operation outweighed the risk to the operators from positive-pressure operation. This view is now supported by almost 250 man-years of data, something that few other projects could provide.

At this stage, we should perhaps look at some numerical values in an attempt to assess the risk issues in cytotoxic work. Consider a typical four-glove isolator of 1 m^3 volume. The isolator is run at 30 Pa positive pressure in turbulent flow at 30 volume changes per hour. As a worst-case scenario, we shall assume that the operator has dropped a one-litre container of a cytotoxic solution and that 1 ml (1 g) has become fully dispersed in the isolator as a result. What are the risks to the operator under these conditions? Possible pathways for the escape of toxic material are as follows:

- The exhaust HEPA filter: Even a HEPA filter provides a potential route for the loss of toxic material, and the designer should be aware of the risk level. The concentration of cytotoxic in the proposed model isolator atmosphere is initially 1 g/m^3, and though this falls exponentially with time, we shall consider that it remains at this level for the first minute. During this time, the exhaust filter is passing 0.50 m^3 of air per minute; thus, it receives a burden of 0.5 g of cytotoxic. Of this, 0.003 percent will pass the filter if the aerosol is sufficiently small, which is unlikely to be the case. However, should this percentage pass the filter, it amounts to 0.015 mg of material leaving the isolator in the first minute.
- Leakage: The most obvious way in which our model isolator may release toxic material is through an unknown leak. Let us assume that the isolator has a hole in the wall of 0.1 mm^2 (0.36-mm diameter). Calculation shows that 0.042 L of air will pass through this hole per

minute and thus, in the initial minute following our accident, the hole will pass 0.042 mg of material. This is about three times the amount passing the exhaust filter. Safety rules suggest that operators should not be subjected to more than 1/1000 of a therapeutic dose, of any drug, during each eight-hour work period. A typical dose of cytotoxic might be 500 ml; thus, exposure to 5 mg is acceptable. This is more than 100 times the dose lost through the supposed leak, at a very high level of contamination. Bearing in mind that the room ventilation will further reduce the room air concentration of the drug in what is hopefully a rare situation, the risk to the operator due to the use of positive pressure in the isolator seems fairly low.

- It is known from the work of the University of Wales (Thomas and Fenton-May 1994) that cytotoxic material can diffuse through virtually all glove materials, creating a problem that the designer cannot solve. The only practical solution here is an operational one — regular glove change according to the SOP for the system.

Now, taking the converse view, what are the risks to the patient if the same isolator model is operated under a negative pressure of 200 Pa? The suggested room environment for such an isolator is European grade D, which permits up to 50 viable organisms per m^3 of air. Calculation then shows that the airflow through our same leak hole of 0.1 mm^2 will draw 3.25 viable particles into the isolator per minute. If the isolator runs continuously, which is probably the case, then, statistically, 546 viable particles will be drawn into the isolator per week. These, of course, will be drawn to the exhaust filter by the ventilation system, be it laminar or turbulent, but some risk of exposure remains. The position of the hole is also very relevant. If it is close to the operator, then more viables would be induced; if it is close to the critical operation, then viables might be directed toward the product. Simple calculation shows that, if the pressure drop across the aperture is 200 Pa, then the velocity of the air through the hole, irrespective of size, is 18 m/sec. Such a jet could carry contamination right into the body of the isolator.

Such figures are purely conjectural, but they suggest that there is a degree of risk to the product and thus to the patient when operating under negative pressure. Whether this is of more consequence than the risk to the operator must remain, to some extent, the choice of the user. Those of a more cynical nature might suggest that the choice is influenced by the likelihood of either party resorting to litigation in the event of mishap.

Turbulent airflow or unidirectional downflow

Following on from the pressure regime, we next need to specify the flow regime and, as described in Chapter 2, we have the choice of turbulent, unidirectional, or semiunidirectional airflow. Once again, we may find ourselves bound less by science than by dogma, because turbulent flow can produce extremely good conditions in terms of the absence of particles, viable or nonviable; but convention dictates that in order to perform certain processes, such as aseptic filling, the airflow should be unidirectional downflow. Experience from isolator processes, such as sterility testing by open membrane filtration, shows that turbulent flow may produce perfectly adequate conditions, but other factors may influence the choice.

In truth, unidirectional downflow does have the real advantage that any particles produced in the isolator are purged out very quickly indeed, which must constitute a reduction in the risk of contamination, but this is balanced by increased complexity in the design, construction, control, and testing of the isolator — which, of course, translates directly into both plant and running costs.

It would seem sensible to adopt a balanced approach to airflow regime in any individual project, taking into account the following:

- The nature of the process: It is likely that only sterile product processes will require unidirectional downflow.
- The pressure regime: If negative pressure has been chosen, then unidirectional flow has some attractions in the potential to reduce the effect of contamination from inward leakage.
- The nature of the material to be handled: Micronised powders, for instance, will produce a much higher particle burden than a liquid product.
- The nature of the process to be carried out: Some processes, such as the capping of vials, tend to produce relatively large numbers of airborne particles.
- The view of the local inspector for the regulatory authority: Some may take the stance that a given process demands unidirectional downflow, whilst others may suggest that if Class 100 conditions can be demonstrated during the process, then the flow regime is irrelevant.

Once again, there seem to be few hard-and-fast rules in this area, and the user is advised to seek opinions from the parties concerned before freezing any given design.

Transfer

As described in Chapter 3, there is a wide choice of transfer ports available, and the writer of a URS may decide to specify some particular type. Alternatively, the URS may set out what is required and allow discussion to develop with isolator manufacturers as to how the problems will be tackled. In order to determine what transfer ports are needed, the following five issues will need to be considered:

1. What items must enter the isolator either before or during the process? There are probably three aspects to this: (1) The product or the main material to be handled: What weight or volume is needed if

this is a batch process? What is the size and the number of containers that will need to enter the isolator? If the process is continuous, then we need to think about how the product is fed to the isolator. (2) We may need to consider empty containers, in particular for filling work. What size are these, and how many are required per batch or per hour? Do they need to be sterile? (3) Do we want to design for change parts, tools, or equipment to enter the isolator during the process, and what provision is to be made for emergency transfer into the isolator when some small item breaks or is forgotten?

2. What products must leave the isolator either during or after the process? In a way, this is just a reversal of inward transfer, but the presentation of the product may well be different at the output stage. Where the process is toxic containment, then design emphasis is going to be placed on the output methods, rather than on the input, where sterile processes are involved.

3. What waste materials are generated? On some occasions, it is necessary to engineer specifically for waste output, and this needs to be considered at the URS stage. An example might be the wastewater or solvent produced during the washdown of a toxic process isolator.

4. What services are required to run the process? Again, these need to be featured in the URS: power, vacuum, steam, nitrogen, and so on — each needs to be cited in the specification to avoid costly modification at a later date.

5. What is the isolator room environment? This is a very fundamental question, especially where sterility is required in the isolator system, and commands the following section of this chapter. Note how various factors, such as airflow regime, isolator pressure, transfer methods, and room environment, interrelate. All of these factors must be balanced when designing an isolator system.

The isolator room environment for sterile operations

James Akers (1995) has suggested that more time has been devoted to the debate about the environment outside the isolator than that inside it. Once again, the lack of fundamental data has led to the application of rather arbitrary decisions by the regulatory authorities, and no real standards exist at this time. Gordon Farquharson (1995) takes a logical standpoint in listing the factors that might be used to decide the appropriate room environment.

- Process risk
- Isolator integrity
- Transfer method
- Isolator pressure
- Sanitisation method
- Regulatory demands
- Operational issues/cost

All of these should be taken into consideration when writing the URS for an isolator project.

The booklet Isolators for Pharmaceutical Applications (Lee and Midcalf 1994) (the *Yellow Guide*) takes a fairly well-considered view and develops a method for choosing the isolator room environment that integrates the process, the isolator pressure, and the transfer method. The argument applies essentially only to processes that require sterility.

The booklet first lists the types of transfer methods generally available and allocates them a letter from A to F based on their quality of transfer, in terms of the mixing of room and isolator atmosphere during transfer. Clearly, simple doors (classified A) on isolators allow much intermixing during transfer, whilst RTPs (classified E) allow no intermixing. If Class A transfer ports are fitted to the isolator, then the room environment must be of very high quality, whilst if Class E or F ports are fitted, the room quality can be much lower.

The booklet then divides isolator processes into the following categories:

- Dispensing — product to be used within 24 hours
- Preparation — product to be used within 7 days
- Manufacture — no specified time limit on product use

Those products to be used in a very short period of time may arguably be prepared under less rigorous isolator room conditions, since there is little time for the growth of any contaminating microorganism.

Next, the booklet sets up a matrix table that places the activity or process (dispensing, etc.) in the isolator along the horizontal axis and the class of transfer port to be used along the vertical axis. A Roman numeral is allocated to each space in the matrix. Low numbers indicate a good-quality isolator room environment; high numbers indicate a poorer quality room environment. For instance, dispensing with Class E ports calls for a grade V room environment, whilst preparation with Class A ports calls for a grade I room.

Two matrix tables are provided, one for positive-pressure isolators and one for negative pressure, because the risk of particles from the room entering the isolator is significantly higher in a negative-pressure regime. Thus, the room standards required for negative-pressure isolators are generally higher than those for positive ones.

Finally, the booklet provides a table that lays out the specific viable and nonviable particle burdens for each of the Roman numeral room environments, laid out alongside the various international standards — the European GMP, the current BS 5295, the previous BS 5295, and U.S. Federal Standard 209E. This integrated approach to isolator design and room environment is still valid but has essentially been superseded by later standards.

In the *Yellow Guide* (1994), the isolator user or URS writer can find some significant guidance regarding the isolator room environment, relating process, transfer ports, and isolator pressure. However, there are some other factors that have a bearing on the issue. Isolator integrity is one such consideration. If it is possible to test the integrity of the isolator easily and frequently (see Chapter 6 on test methods), then a lower-standard room may be used than would be the case if leak testing is only rarely possible. Another consideration is the method of sterilisation or sanitisation. If simple alcohol spraying is employed, then a better room environment will be indicated as compared with, for instance, gas-phase sterilisation. Finally, there may be operational considerations, such as the product value and the level of operator training.

Having looked at all of these factors, however, a consensus has developed that suggests that general sterile process isolators should be placed in a room environment equivalent to European grade D, now Class 7 in the ISO BS EN 14644. If any factors make the isolator system more critical, then this should be raised to grade C (ISO Class 6). The British GMP (1997) goes so far as to directly specify grade D for the housing of isolators.

Where the isolator application is purely one of toxic hazard containment, then room conditions are probably much less critical. Consideration would probably centre on the air change rate in the room, and room pressure with respect to the exterior, creating a cascade of depression from corridor to room and from room to isolator.

Control and instrumentation

Control and instrumentation will also need to be addressed in the URS, to suit the process in hand. As discussed in Chapter 4, control may be manual or automatic, whilst instrumentation may consist of simple analogue instruments or complex digital monitors with process interlocks, alarm systems, and data handling. The control system should at least allow the user to set and maintain a certain isolator pressure and airflow rate, which may then be validated, if required.

At a minimum, there should be instrumentation of the isolator pressure and the airflow rate, whether turbulent or unidirectional. Ideally, these parameters should be alarmed for both high and low excursions outside user set and validated limits. This alarm needs to be audible, visual, and possibly remote, and it may be process integrated. For example, a filling machine might be arrested automatically if the isolator pressure falls outside a set limit. Alarms should also be latched: that is, the alarm signal should remain in place until reset by an operator, even if the isolator has returned to normal working conditions. The issue of alarm latching is really quite fundamental to isolator operation and seems to be frequently overlooked by manufacturers.

Beyond these minimum control and instrumentation requirements, the process may dictate other parameters, such as the need for continuous particle counting or humidity and temperature measurement. The degree of sophistication may, in the final analysis, be limited by the available funds.

A particular point to note here is that some process isolators are sited in production areas that handle solvents and are, therefore, classed as Zone

1 hazardous. In this case, the entire electrical system on the isolator, including instrumentation, will need to be Zone 1 flameproof to the required standard, a factor that must be addressed in the URS since it may affect costs considerably.

If possible, P&IDs should be provided in the URS for process isolators, so that the necessary equipment can be integrated into the design. These should give as much information as possible, including dimensions of the proposed pipe and ductwork; flow rates; pressures; valve types and specifications; and the placement of heating coils, cooling coils, dehumidifiers, and the like. Indeed, the P&ID may become an integral part of the isolator design in many cases. The early provision of function specifications that apply to the P&ID will also help in the design of the control system.

Sterilisation and decontamination

Sterilisation and decontamination issues are dealt with in some detail in Chapter 7, but the sterilisation of aseptic isolators and the decontamination of toxic containment isolators must be clearly addressed in the URS. For sterile isolators, this may mean the provision of sterilising gas input and exhaust points; for toxic isolators, it may mean the fitting of a water supply and a waste drain valve, such as an Asepco Radial Diaphragm Valve ™.

Recent isolator projects are specifying CIP/SIP for production isolators, and this will mean the provision of carefully designed spray systems and engineered draining arrangements, perhaps combined with a warm air drying facility, and sterilising gas input and exhaust points. Such systems need to be integrated into the isolator design from the outset and must be mentioned in the URS if they are to be fitted.

Standard versus special isolator design

The question of standard versus special design has been something of a problem for isolator manufacturers. Most will offer a range of standard designs — two-glove, four-glove, positive, negative — and also a special design and construction service for particular applications. Naturally, a specially designed isolator system will cost much more than an existing standard design, and there is often pressure on the manufacturer to modify a standard design rather than go to the expense of custom production. Perhaps because of the lack of standards and guidelines for isolator design, manufacturers have been willing to do this, but the result may not be very effective and can be surprisingly expensive to produce.

Few isolator users will realise that it may actually increase the cost of a standard isolator to remove a part, such as a transfer port. As an example, let us say that a simple door is to be taken off a standard flexible film glove isolator. The support frame and canopy now have to be redrawn and then sent for special manufacture, since they cannot be taken from standard stock. The bill of materials must be amended with new part numbers for the modified parts and the deletion of those parts not now required, which then become surplus stock since they do not now match up with standard unit inventory.

The effect of this situation is either to leave the manufacturers with an insufficient profit margin or to leave the user with the impression that he or she has been charged very heavily for what appears to be relatively trivial modifications. Neither situation is desirable, and it is important for both parties to reach an understanding of standard and special designs at an early stage in the project cycle. Figure 5.2 and Figure 5.3 show isolators resulting from careful URS work. Figure 5.2 is a process containment suite and Figure 5.3 is a powder-handling isolator.

Figure 5.2 A Special Stainless Steel Containment Isolator Suite Built by Envair Ltd.

Figure 5.3 Powder-Handling Isolator. The lockchamber in the foreground is used to load kegs of product; note the rollers to ease handling. (Courtesy of Envair Ltd.)

chapter six

Seeing the project through

Where established standard equipment is involved, the process of liaison with the supplier should be relatively simple. The manufacturer's sales or applications engineer will be able to advise on the optimum equipment for the work to be undertaken. There is a lot of precedence for deciding on isolator pressures, flow regime, transfer, sterilisation, room conditions, and the like. This would be the case, for instance, in hospital pharmacy work or in sterility testing. Even so, as we have seen, isolation technology is still to some degree in its infancy, and the project manager would be well advised to take into consideration many of the aspects discussed in this book. This would allow the manager to confirm that any system proposed by the supplier meets the requirements of the project in terms not only of the process to be carried out but also of ergonomics and safety, documentation and validation, etc.

If the project involves specially designed isolator systems to suit a particular process, then liaison with the supplier becomes a critical process, by which the project may sail through to successful validation or sink after striking the iceberg of misunderstanding. The communication process becomes even more critical if other equipment suppliers are closely involved, as in the case of isolated filling lines. Even more complications arise if, as often happens, the customer, the isolator manufacturer, and the equipment manufacturer are all in different countries. The scope for failure to appreciate the situation in each group, with consequent divergence of action, is all too wide in this situation; thus, a very professional approach will be needed by all those involved.

In cases where multiple manufacturers are involved, many clients will prefer to set up a turnkey operation, with one of the manufacturers becoming the prime supplier or turnkey manager and handling the overall management of the project. The order for the project is then placed with the prime supplier, and the other manufacturers then act as subcontractors. Clearly, it is an advantage to the client if only one company is dealt with for the greater part of the project. In some cases, the turnkey company is a building contractor with overall control, literally from the ground up; in other cases, the turnkey will be whichever company has the largest cost share of the project. However, there is a good case for the isolator manufacturer to be the turnkey operator, since the isolators will influence not only the design of the process equipment but also the design of the isolator room and its environment.

Liaison with your designer and supplier

Put in very simple terms, communication must be established so that the isolator manufacturer can understand what the client wants to achieve and, at the same time, the client can understand how the manufacturer plans to reach this goal. The volume of information that must be transmitted is, however, quite large, as illustrated in Chapter 5 of this book, and the sooner clear lines of communication are established, the better.

Whilst there are various advanced forms of data exchange, the fundamental basis of communication will always be person-to-person meetings across a table. It is probably a good plan to put forward one person from the client team and one from the isolator supplier team to form a focus for communication. These could well be the project managers from each side, and ideally they should have not only engineering capability but also some degree of commercial experience. They, in turn, will be supported by project engineers, QA managers, commercial managers, designers, and so on. At times, these people may talk directly to one another, but when major decisions need to be made, or when problems arise, the two designated individuals need to be in control. Hopefully, the two team leaders will establish a good rapport that will allow them to exchange ideas and concepts and to discuss management or commercial issues freely and candidly.

The team leaders should then be responsible for setting up lines of communication and for establishing who is responsible for each area of activity in their teams. This would include the obvious aspects of telephone numbers (including mobiles), fax numbers, and e-mail addresses. In addition, it is often very helpful if the two design teams can exchange computer-aided design (CAD) drawings by modem, so software engineers should be involved to ensure compatibility. Both sides might produce a management plan to ensure that all aspects of the project have personnel assigned to tackle them. This might take the form of a block diagram or a written text describing the management structure. Following this, a broad project plan might be established laying out dates for formal progress meetings at, say, monthly intervals, perhaps alternating between the client site and the manufacturer site, wherever practical. Less formal progress meetings might take place weekly; but in both cases, if these meetings are to be effective, they should have an agenda, and the discussions must be reported and distributed to all participants. Apart from the question of good management practice, documentation is an integral part of any pharmaceutical installation, and the documents recording the development of a major project could be directly relevant to the final documentation package.

Further project development

The project plan

The next stage is to develop a detailed project plan, typically summarised in the form of a Gantt chart, which will establish the time schedule of the project, including the vital milestones, such as design approval and factory acceptance test (FAT): a sort of predelivery inspection. This, in many ways, becomes the key document, as it is the yardstick by which progress may be judged. Whilst it may change as the project moves on, this project plan should be established at an early stage and then monitored closely through-out the period of the project. Each Gantt chart will, of course, be tailored to its individual project, but typical stages in the chart might be as follows:

- URS approval
- General design
- Detail design
- Design approval (DQ)
- Fabrication
- Parts procurement
- Assembly
- Test
- Predelivery inspection and FAT
- Packing and shipment
- Assembly and installation
- Site Acceptance Tests (SAT) and IQ
- Commissioning
- OQ
- Preproduction work (PQ)

DQ, IQ, OQ, and PQ

The concept of qualification documents was introduced at the start of Chapter 5; these are the design, installation, operational, and performance qualifications that may be defined briefly as follows:

DQ — design qualification
This was discussed in Chapter 5.

IQ — installation qualification
The IQ document is often written by the isolator supply company, following detailed discussions with the client, though it may be written by the users or by outside validation contractors. The IQ simply lays out how and where the isolator will be installed.

OQ — operational qualification

The OQ document is, again, often written by the supply company, following discussion with the client, but may be written by others. It lays out what operational tests will be carried out on the system once it has been installed, what the conditions of the tests will be, and the criteria of acceptance of those tests. It essentially describes the commissioning work to be carried out. The FDA-preferred format for the OQ document is as follows:

- The name of the test
- The purpose of the test
- The equipment required
- The method proposed for testing (method statement)
- The criterion (or criteria) of acceptance
- Table of results
- Pass/fail statement
- Comment space
- Signatures of the test engineer and client witness, time, date, and place
- Calibration certificates for test instrumentation used
- Raw data file (if applicable)

The OQ document can become fairly lengthy for larger isolator systems and can be a rather repetitive piece of literature, but it does form a central anchor point from which the performance of the system can be judged by all concerned. This does not, however, excuse recent examples of isolator OQs, which are not only exceedingly lengthy but laid out in a confusing and unhelpful manner. When developing OQ protocols, writers are urged to adhere to the format laid out above and to keep each single OQ check confined to consecutive pages in the protocol. Placing method statements in one section of the protocol, tables of results in another, and pass/fail statements in yet another section makes the protocol difficult to read. The OQ should be written at an early stage in the project and may form part of a preengineering study (see "Preengineering Studies, Models, and Mockups"). Typically, the type of physical tests described in the OQ will be as follows:

- HEPA filter DOP tests
- Isolator leak tests
- Instrument calibration
- Airflow pattern check
- Particle counting and possibly particle recovery
- Relative humidity and temperature checks
- CIP drainage and drying tests
- Alarm tests
- Breach velocity check
- Interlock check

The OQ is discussed in more detail later in this chapter. Figure 6.1 is a typical OQ test sheet for instrument calibration.

1. Name of Test.
   Calibration of the isolator pressure meter
2. Purpose of the Test
   To calibrate the isolator pressure meter against a traceable test instrument
3. Equipment Required
   Calibrated micromanometer, hand air pump, tee-connectors, tubing
4. Method Statement
   a. Disconnect the tube connecting the isolator pressure meter to the solator chamber.
   b. Connect onto the isolator pressure meter the test instrument and the hand air pump.
   c. Raise the system to a series of test pressures using the hand air pump and record the readings of the test instrument and the isolator pressure meter in the table below.
   d. Return the isolator to normal operational condition
5. Criterion of Acceptance
   The isolator pressure meter shall read within plus or minus 10% of the reading on the test instrument across the range from zero to 250 Pa
6. Table of Results

Test Instrument Pa	0	50	100	150	200	250
Isolator Pressure Meter Pa						
Test Instrument Pa	250	200	150	100	50	0
Isolator Pressure Meter Pa						

7. Pass / Fail Statement

The isolator pressure meter reads within + or − 10% of the reading on the test instrument across the range from zero to 250 Pa (Yes or No)	

8. Comment Space

9. Verification

	Name (Print)	Company	Sign	Date
Test Engineer				
Client Witness				

10. Calibration Certificates

Calibration Certificate for the test instrument is attached (Yes or No)	

11. Raw Data

Raw data for this test is attached (Yes or No)	

Figure 6.1(a) and (b) An Example of How a Typical OQ Test May Be Laid Out, in This Case for Instrument Calibration.

PQ — performance qualification

The PQ document is generally written by the client company, with help from the supply company, if required. In a similar format to the OQ, it describes how the performance of the completed system of isolator and process equipment will be measured and thus judged to fulfil its overall design requirements. Typically, this might include the following:

- Functional test of the CIP process
- Microbiological challenge of the sterilisation process
- Functional test of any decontamination process for toxic applications
- Ergonomic and safety function tests
- Media filling trials
- Particle counting during operation
- Viable particle testing during operation
- Stability of control during operation
- Alarm function during operation
- Mechanical and electronic reliability

The PQ is discussed in more detail in Chapter 8.

Some further points about qualification protocols

- It should be noted that validation must take place sequentially, and thus no protocol can be executed until the previous one has been approved. Originally, this simply meant that the protocol was signed off by the designated parties, but current practice requires that a separate protocol report be written and approved. Only when this report has been approved can the next stage be executed.
- Given this requirement for approvals, protocol writers should take care to minimise the number of approval signatures required. Project managers, production managers, technical directors, QC managers, and validation managers tend to be very busy people. Experience indicates that obtaining approval signature from these key workers can be a lengthy and frustrating business. By keeping protocol signatories to a minimum, the delay before starting the next execution can be reduced.
- Another issue which can trip up the validation process is that of Standard Operating Procedures (SOPs). These are required to be approved before the OQ can be completed, though approved draft SOPs seem to be acceptable in this context.
- Other items to address at this stage include plans for maintenance (Planned Preventative Maintenance — PPM) and instrument listings.

Proposals and quotations

The scope and volume of any proposal put forward for an isolator project will depend on the degree of specialisation that the isolator project requires. For standard isolators, the proposal may amount to a very simple check that the suggested isolator is suitable for the process; this might be the responsibility of the isolator sales engineer. Where more complex systems are involved, then settling as many issues as possible — technical, commercial, and logistic — before metal is cut will save time and money in the long run. In some cases, it is normal for the client company to issue a specification, essentially the URS, and for several manufacturers to make proposals and quotations against this specification. Having written a good URS as described in Chapter 5, the client requiring a fairly complex and sophisticated isolator system may expect to receive a proposal from the manufacturer containing documents such as the following:

- Company information: number of employees, current turnover, factory size and facilities, BS or ISO accreditation, and relevant experience level.
- A general specification for the proposed isolator, including the following paragraph headings:

  Subject
  Scope
  Definitions
  General description of the system
  Materials and methods of construction
  Ventilation and air handling
  Control
  Instrumentation
  CIP
  Sterilisation
  Transfer methods
  Handling methods
  Factory acceptance, packing, transport, and installation

- A functional specification for the proposed isolator, including the following paragraph headings:

  Scope
  Air filtration quality
  Airflow rates
  Air pressure regimes
  Air cooling and heating requirements
  Containment level
  Room conditions
  Ergonomics and safety

- A control specification. In some cases, the manufacturer may wish to write this only after an order has been placed, since it may be very project-specific.
- A project plan, such as a Gantt chart.
- A management plan.
- A quality plan.
- Outline or draft IQ and OQ.
- Drawings, including three-dimensional general arrangements and P&IDs.
- The quotation, preferably not as a lump sum, but broken down to show the price of the major items, any options, spares package, delivery, commissioning, and any discount applicable.
- A reference list, with contact names and telephone numbers.

Prospective isolator purchasers should bear in mind that the preparation of this kind of documentation can be a lengthy process and commits the isolator supplier to considerable effort with no guarantee of a successful bid. Thus, a period of one week from the issue of a tender to the deadline for bid submission is not sufficient; at least one month should be allowed. During this time, the purchaser should expect to answer a number of queries as the supplier develops the proposal; if there are no questions, the supplier probably does not understand the tender.

Preengineering studies, models, and mockups

As suggested above, some isolator systems are now so complex that it is hard for the manufacturer to quote accurately without carrying out a fair amount of development work beforehand. In this case, the client may choose to place an order for a preengineering study. This will establish the basic design parameters for the system and allow the manufacturer to quote a fair price, after which the client may proceed with the full order and deduct the cost of the preengineering from the final total, or the client may simply pay for the study and drop the project.

An important part of the preengineering phase may be the development of a computer model in order to test the ergonomics of a process, at a minimum cost if possible. Most CAD packages have some form of human ergonomic capability, and more specialised programs are available with the capacity to draw figures of a particular stature. The movement of these figures can then be drawn out in a series of steps to show how, for instance, a transfer port may be reached by a half-suit operator.

The use of computer modelling will go a long way toward optimising design, but, in some cases, a full-size mockup may be a worthwhile investment before finalising design. Timber, Dexion™, plywood, Medium Density Fiber (MDF), and plastic are all pressed into service to make a realistic but cheap ergonomic mockup, which could be used, for instance, to establish the best possible position for gloves to access a piece of process equipment.

Another aspect that should be considered at the preengineering phase is the practicality of delivery of the finished units to their final position on-site. If access is easy, on the ground floor, with wide doors opening far enough for a forklift truck to enter, then site placement will be simple. If, on the other hand, access is limited, up stairs, or through narrow corridors, then the isolators may have to be delivered in kit form for assembly on-site. This is no great problem for flexible film construction, but if the isolator is a rigid stainless steel unit, installation could be difficult.

Terms, payment structures, and guarantees

Whilst not strictly part of a practical guide to isolation technology, the commercial aspects of any project should be addressed at an early stage, as there is little sense in developing a good technical solution, only to find that the form of payment requested by the manufacturer cannot be accepted by the client. This is not likely to be a problem in the case of standard isolators in small quantities, but if the project is relatively large, i.e., more than £50,000, then the terms of payment must be agreed upon early on in the discussion. With special projects, such as filling lines, there could be a time interval of more than a year from placement of order to completion of commissioning. It is unreasonable to ask even a large and established manufacturer to support the work entirely for this length of time. So, the normal practice is to set up a payment schedule; a typical example might be as follows:

- 20 percent on placement of order
- 20 percent on delivery of major items, such as stainless steel shells
- 30 percent on satisfactory predelivery inspection
- 20 percent on completion of on-site commissioning
- 10 percent retention (performance bond) for one year following delivery

Finally, the precise form and duration of the guarantee and/or performance bond should be agreed upon prior to order, to avoid potential conflict at a later date. Time spent in careful groundwork of this nature will speed the project through to a satisfactory conclusion.

The OQ in more detail

The OQ was mentioned briefly earlier in this chapter, but its importance to any isolator project is such that it demands further attention. The OQ is a fundamental document for any piece of pharmaceutical equipment, including both standard and special isolators. Where standard units are concerned, the manufacturer will no doubt offer its own in-house test protocol that may be repeated on-site following installation. At a minimum, this will probably consist of HEPA filter testing, leak testing, and possibly instrument calibration — these being the main parameters that will dictate the air quality within the isolator working area or critical zone. In standard isolators, factors such as airflow pattern should be well established and need not be examined as part of the OQ.

Special isolators, however, may involve quite lengthy OQ documentation, addressing a number of parameters in more or less detail, according to the specific requirements of the project. In some cases, the OQ work overlaps with the PQ work. Generally speaking, the tests would be as follows.

HEPA/ULPA filter test

The isolator is very much reliant on the efficiency and integrity of its inlet air filters and, to some degree, on its exhaust air filters for maintaining required air quality within the working area. Occasionally, ULPA filters will be specified, but as a rule, the inlet and exhaust filters will be HEPA standard — each one must be individually tested as such. This will involve the use of a smoke generator to produce a suitable challenge on the upstream side of the filter and a measuring device, such as a photometer, to check the resulting particle challenge on the upstream side of the filter and then measure the burden immediately afterward on the downstream side.

The two methods of testing are Volumetric and Scan. The Volumetric Test method gives an overall efficiency value of the sample challenge and will not identify the leakage point. This is usually directly related to the filter efficiency, i.e., 99.997 percent leakage penetration of 0.003 percent. The Scan Test method is carried out approximately 20 to 30 mm from the filter face and seal and will indicate positions of leak on the media and seal. This does not usually relate to the efficiency of the filter but conforms to standards, i.e., BSPD6609–2000 or BS EN ISO 14644–3 (when finally issued): a maximum leak penetration of 0.01 percent on a 99.997 percent rated filter.

Smoke generators produce an aerosol with a particle size around 0.50 μm from a suitable oil. Until recently, the oil was dioctyl phthalate, hence the generic name DOP test, but since there are doubts about the safety of this substance, a food-grade light oil (e.g., Shell Ondina EL or Emery 3004) is now used, and the acronym is redesignated *dispersed oil particulate*.

There is some controversy as to whether the smoke generator should be of the hot or cold smoke type, but in either case, the test will probably be based on BS 5295, Part 1, Appendix C, giving a smoke challenge of between 50 and 100 mg/m^3. Where two filters are fitted in line, such as the exhaust side of cytotoxic isolators, it is necessary to test each filter individually. Hopefully, the isolator will be engineered to make this possible (and will be defined in the DQ).

Panel filters, such as those in the roof of unidirectional downflow isolators, should be scanned with the photometer. This cannot be done with most exhaust panel filters and also the canister-type filters often found on smaller isolators. For these cases, only an overall penetration value (volumetric) can be obtained (see Chapter 2, "Filtration").

Leak testing

The isolator also relies on the integrity of its physical structure for maintaining the internal air quality. There seems to be more debate about the leak testing of isolators than any other aspect of their performance, almost on par with the debate about room conditions. First of all, there is the philosophical debate as to whether leak testing is needed at all. Any leakage in a positive-pressure isolator for sterile application will be outward, protecting the product; the converse is true for negative-pressure containment in toxic applications.

Leaving aside the case of combined sterile and toxic work, the argument may be basically sound, but, in practical terms, it is surely necessary to establish some kind of standard by which to judge the integrity of the isolator structure. One school of thought suggests that it would be appropriate if the leakiness of the isolator were roughly on par with the penetration of the air filters, since these two parameters have the major influence on internal air quality. The relevance of leakage varies according to the type and use of the isolator. In positive-pressure aseptic isolators, leak-tightness (or "arimosis") is important when gas sanitisation is used. Clearly, escape of the gas could be hazardous to the operators. In negative-pressure aseptic isolators (e.g., for cytotoxic applications), leakage is very significant, since contaminated air from the room will be drawn into the isolator. At only moderate negative pressures, the velocity of the inflowing air jet will be several metres per second, thus penetrating deep into the critical workspace. Having agreed that leak rate measurement is part and parcel of isolator operation, we then have to devise methods and criteria that can provide the necessary data, and we need to match these to the type and use of the isolator so that appropriate standards can be applied.

At this stage, it is important to distinguish between leak test methods which make a measurement of the leak rate and those which locate, but do not quantify, leaks. **The prime target of leak testing should be to quantify the leak and thus establish if the isolator has an acceptable or an unacceptable rate of leakage**. If excessive leakage is evident, then the source of leakage must be located before it can be repaired. The methods used for expressing leak rate are discussed further on in this chapter.

Most test methods will involve taking the isolator to a test pressure that is more than the normal operating pressure, and suggestions range from 1.5 to 5 times operating pressure. Care should be exercised because excessive pressures can damage the structure of an isolator, and, perhaps curiously, it is rigid isolators that will be damaged by excess pressure, while flexible film isolators can withstand high pressures, both positive and negative. The manufacturer must be consulted, but as a rule of thumb, rigid isolators should not be pressed to more than 2 times operating pressure and flexible film isolators not more than 5 times operating pressure (to be defined in the original design poroposal).

It is fair to say that all leak test methods are affected by changes in the internal temperature of the isolator and changes in atmospheric pressure during the test. If the test itself is not directly affected (e.g., the Parjo test — see below) then the true rate of leakage will be still affected by these changes. Short tests are thus preferred in order to minimise the effects of these changes, whilst in some cases, correction can be made to allow for changes in internal temperature and atmospheric pressure.

Leak detection methods

Soap test

In terms of equipment required, the soap test as described in BS 5726, which relates to biological safety cabinets, is probably second to none. However, it is a rather messy process, labour-intensive, operator sensitive, and likely to be relegated to only the smaller and simpler type of isolator. It is also only appropriate on isolators that can be positively pressurised, though it can be used whilst the isolator is in operation.

Gas leak detection tests

Gas leak detection can conceivably be used in neutral pressure, but like most leak tests, these normally rely on being able to seal the isolator by closing off the inlet and outlet valves, or by fitting sealing plates to the air inlet and exhaust, and then raising the internal pressure to some level above the normal operating pressure. The suggested test pressure has been put at anywhere between 1.5 times the working pressure (*Yellow Guide*, Lee and Midcalf 1994) and 5 times working pressure (Advisory Committee on Dangerous Pathogens (ACDP) test for flexible film isolators, 1985).

Perhaps the simplest and still one of the most useful, the gas test uses ammonia, detected on the outside of the isolator by bromophenol cloth, which changes from yellow to blue in the presence of the gas. A few millilitres of ammonia solution are poured into a petri dish inside the sealed and pressurised isolator. Then the operator slowly passes a patch of bromophenol cloth across the surface of the isolator, along all the welds, and around the joints and seals, examining the cloth frequently for signs of colour change.

This method is sensitive and can pick up quite small leaks, but is, like many methods, entirely subjective. It leads to the prime question: What is a significant leak? The depth of colour change is influenced by the concentration of the gas, the time of exposure, and the size of the leak. The method cannot be standardised in such a way that will give a repeatable numerical value to the integrity of the isolator. It is also a particularly time-consuming, labour-intensive method, demanding care and patience from the operator if meaningful results are to be obtained. Thus, it is open to misinterpretation.

Helium passing through a small aperture can be detected with a gas thermal conductivity meter, whilst (environmentally acceptable grade) freon can be detected using an electronic device similar in operation to a Geiger counter. Thus, one of these gases may be introduced into the sealed and pressurised isolator, and then the outside can be scanned for leaks. Again, this method is very subjective and relies heavily on the care of the operator to test every possible leak path on the isolator. However, it does have a practical advantage over ammonia in that the instruments will quickly give a particular signal in response to a given leak. This makes it fairly easy to return to a known leak site and check the response at any time. In this way, remedial action can be taken where a leak is found and the resulting improvement quickly demonstrated. This is very useful in the fine-tuning of an isolator, but, again, it does not give an overall figure for the integrity of the system. Gas detection is still, however, widely used in routine leak testing.

DOP smoke tests

DOP tests require the same equipment as the HEPA filter tests described earlier: a smoke generator and a photometer. The isolator is filled with smoke and the photometer is used to trace leakage. This is a method of high sensitivity and one which reportedly defines leaks clearly, unlike gas tests, which give readings that are more difficult to interpret.

This test has a noticeable advantage over most others in that the isolator need not necessarily be sealed and pressurised; it can be used on negative-pressure isolators and on some of the unidirectional flow isolators that have panel filters as the air inlet and exhaust. It is still helpful if the isolator can be pressurised, since this will provide a greater challenge, but smoke will nonetheless diffuse to a measurable extent through leaks in neutral pressure. Some workers test negative-pressure isolators by placing the photometer head inside the isolator and wafting smoke along the outside of the seals and joints to be tested (Lumsden 1996).

A significant point in favour of the test is that the smoke particles are of similar dimensions to the microorganisms that we wish to exclude from sterile isolators; thus, the challenge is a representative one. If the isolator is leak-tight to DOP smoke, then it is arguably leak-tight to microorganisms, whereas an isolator that leaks significantly to a pressure hold test might yet be microbiologically sound.

It is also possible to put some sort of figures to this test, making it less subjective and more repeatable than others. Indeed, we can set up the smoke challenge and the photometer response (full-scale deflection at 100 µg/l or mg/m^3) in the same way as we have done for the HEPA filter test, and thus quantify any leaks with a penetration value. However, we are still left with the question as to what is an acceptable, overall leakage value.

A minor disadvantage of the smoke test is that it coats the internal surfaces of the isolator with oil that must eventually be cleaned off; it also puts a burden on the HEPA filters that may shorten their working life. It is still a convenient test to run if the HEPA filters are being checked at the same time. This test is widely used and accepted, particularly in the pharmaceutical industry.

Pressure tests

Pressure decay

The principle of the pressure test is very simple: Pressurise the isolator to some suitable level, seal the inlet and exhaust, and monitor the pressure for a period of time. If the isolator leaks, the pressure will decay; if it is leak-tight, then the pressure will hold steady. There are variations, but the principle remains the same. In practice, the test is much more complex, but the attractions of the method are such that it is well worth considering in detail. Pressure testing can be applied to negative as well as to positive isolators, but, clearly, it must be possible to seal the system completely.

The major advantage of the pressure decay test is that, with certain major provisos, it can provide a repeatable, quantitative, nonoperator-sensitive leak test. If the test pressure and the time period are fixed, then the pressure decay will give a figure that can be used in its own right as measure of isolator integrity, or it can be used to calculate some other comparative measure, such as the percentage volume lost. This really is the sort of data we are seeking when leak testing an isolator. If it is assumed that the loss of pressure is due to a single aperture, then it is even possible to calculate the effective diameter of this hole and then decide if it is operationally significant or not.

The major disadvantage of the pressure decay test is that it is very sensitive to the conditions of the test, specifically, the stability of the isolator internal temperature and atmospheric pressure, during the period of the test. The Universal Gas Law governs the relationship between the pressure, volume, and temperature of a fixed mass of gas where P = absolute pressure (Pascals), V = volume (cubic metres), and T = absolute temperature (degrees Kelvin):

$$\frac{P_1 V_1}{T_1} = \frac{P_2 V_2}{T_2}$$

Thus, if the temperature inside a sealed isolator changes from 20 to 21°C, then the isolator pressure will increase by nearly 350 Pa. Similarly, if the atmospheric pressure changes by 1 mb, then the isolator pressure will change by 100 Pa. Since atmospheric pressure can change by 10 mb/h at times, the potential for error is large. Beyond this, the volume of an isolator is not fixed, even if it is a rigid structure, because of the presence of sleeves or half-suits, which are very flexible. A change in volume of 1 percent, caused, for instance, by the movement of a sleeve, changes the isolator pressure by 1000 Pa.

The corrections for internal temperature change and atmospheric change during a pressure decay test, at either positive or negative pressure, may be given as follows:

- Internal Temperature: If the internal temperature has gone up, subtract 3.5 Pa from the final pressure reading for every 0.01°C of change. If the temperature has gone down, add 3.5 Pa.
- Atmospheric Pressure: If atmospheric pressure has gone up, add 1 Pa to the final pressure reading for every 0.01 mb of change. If the pressure has gone down, subtract 1 Pa.

Clearly, the effects of internal temperature and atmospheric pressure change constitute very serious drawbacks for the pressure-decay method, and they must be carefully considered in applying the test. Indeed, the original *Yellow Guide* goes so far as to state: "The pressure decay test can only be relied upon to give indication of a gross leak." (Lee and Midcalf 1994) This is a rather extreme view and, in practice, the test can be used very successfully. The first palliative action is, of course, to minimise the length of the test and thus minimise the effects of environmental variation. There is no standard here, but a period of a few minutes seems too little to establish a decay curve, and a period of some hours seems too prone to outside influence. A time frame in the region of 10 to 30 minutes seems practical, and most isolator rooms hold a surprisingly stable temperature for this length of time, if sensible precautions are taken. Clearly, all local heating should be avoided. Testing early or late in the working day should also be avoided, as the building's own heating may be ramping up or down. If the room has windows, these should be covered to prevent sunlight acting on the isolator, directly or even indirectly, thus heating its internal atmosphere. Room temperature and atmospheric pressure should be monitored during the test and the results related to these. For basic leak tests (e.g., unidirectional flow isolators), it is sufficient to monitor external temperature and atmospheric pressure and accept the raw pressure decay figures, provided that temperature and pressure have not changed significantly. For more rigorous tests (e.g., Class 3 safety cabinets), the internal temperature and atmospheric pressure can be measured at the start and end of the test and the mathematical correction given above applied.

Given stable conditions and a test period of, say, 15 minutes, what test pressure should be applied and what section of the resulting asymptotic decay curve should be used as the actual test data? As previously mentioned, test pressures from 1.5 to 5 times working pressure have been suggested — perhaps twice working pressure is a fair challenge. Regarding the starting point of the test records, the ACDP guidance for flexible film isolators (1985) requires that the isolator be inflated to test pressure and sealed for 30 minutes, then reinflated to the test pressure and tested for a further 30 minutes. With rigid isolators, this dwell period might be rather less, since the wall structure is not elastic. The ACDP then allows a pressure decay of up to 10 percent within the test period, which seems a reasonable criterion of acceptance for the average isolator application.

It has been pointed out that flexible film isolators are, to some extent, elastic and will thus hold the test pressure for a longer time than an equivalent rigid structure. This may be true, but plots of pressure decay against time do not generally show a level period followed by increasing decay. Provided that the same test conditions are applied each time, pressure decay still seems a good yardstick by which to judge the "arimosis" of flexible film isolators. A further complicating factor, and one that applies to rigid and flexible isolators alike during pressure decay testing, is the matter of sleeves and half-suits. Being very flexible, these can move and alter the effective volume, which, as previously mentioned, can create a big change in pressure. The convention with sleeves is to evert them fully (i.e., pull them out of the isolator), together with the gloves, during pressure testing. They will then act as pressure compensators to some extent, but the effect does not seem to detract from the repeatability of pressure decay tests. As regards half-suits, the best action seems to be to settle them into a stable position by gentle agitation at the test pressure, and to avoid disturbing them subsequently. The suits must be suspended by rigid ties and not elastic ropes during pressure testing, since the elastic will make the suits act as major pressure compensators.

Pressure hold

Another variation of pressure testing is to seal the isolator and then pump in air to maintain a given test pressure. Measurement of the flow rate required to maintain pressure is then a direct measure of leakage at that pressure. This is a simple and effective test but is really only practical for isolators that may be described as relatively leaky, giving perhaps more than 1.5 percent volume loss per hour. In this case, a centrifugal fan controlled by a lamp dimmer and a ball-in-tube type flow meter (e.g., Rotameter) can be used to quantify leak rate.

Parjo

This fairly sophisticated pressure test for gloveboxes is described in the AECP 1062 (1981) and more recently in ISO BS EN 14644–7, and is known as the Parjo method, after the inventors K. Parkinson and W.F. Jones. In essence, a small air reservoir, such as a Winchester bottle, is placed inside the isolator. The top of this reservoir carries a glass tube with a horizontal section containing a soap-bubble meniscus, which acts as a frictionless piston. Any change in the pressure of the isolator will cause the meniscus to move along the tube: given the diameter of the tube and the velocity of the meniscus, a percentage leak rate can be easily calculated. This system takes into account the environmental variation to a certain extent because, in theory at least, the small reservoir will respond to temperature change at the same rate as the main isolator. Atmospheric pressure changes do not alter the differential between the reservoir and the isolator. Thus, the method could form the basis of a standardised pharmaceutical isolator leak test, but, up to this point, it has not been significantly used for this purpose. Perhaps the main drawback of the method is its extreme sensitivity, such that the average isolator will eject the soap bubble before a measurement of speed can be made. The suggestion has been made that the soap bubble could be replaced with an electronic micromanometer reading the differential pressure between the Parjo vessel and the isolator. This method has been christened the Fosco test after the inventors, M. Foster and the author. Some basic research and development of this test method might be well received by the industry.

The oxygen test

This method is well described in ISO BS EN 14644–7. In this case, an oxygen meter is placed inside the isolator, which is then filled with nitrogen and held under a negative pressure. The rate of increase in oxygen within the isolator is a direct measure of the leak rate. This is a very sound method but may not be very practical for the average user of isolators, the test being better suited to type testing of standard designs.

Expression of pressure decay test results

Having carried out a pressure decay leak rate measurement, the results may be reported in a variety of ways.

Hourly leak rate

This is the expression of choice in ISO 14644–7 and also ISO 10648–2, and is defined as the ratio between the hourly leakage of the isolator (at the test pressure) and its volume. This is then expressed in reciprocal hours (h^{-1}). The formula for calculating this figure is very simple:

$$\text{Hourly leak rate} = P1 - P2/P2$$

Percentage volume loss per hour

This is the volume of air lost (or gained) from the isolator per hour, at the test pressure, expressed as a percentage of the volume of the isolator. It is simply the hourly leak rate expressed as a percentage. This form of reporting is favoured by many as being easily understood, though it is not accepted by all parties.

Volume loss per second (or per hour)

This is simply the volume of air lost from the isolator per second, at the test pressure.

The formula for calculating this is given as:

$$\text{Volume loss per second} = V \times \text{hourly leak rate}/3600 \text{ m}^3/\text{sec}$$

Single-hole equivalent

This is the diameter of a single hole, which would account for all of the observed leakage. A number of assumptions are made in calculating this value, which in any event is a very theoretical concept. It is nonetheless a useful tool in quantifying the "arimosis" of an isolator.

Classification by leak rate

ISOs 14644–7 and 10648–2 effectively provide three classes of isolator according to their leak rate. Table 6.1 provides comparison of the classes with the methods of reporting. Note that in the case of volume loss per second and single-hole equivalent, the figures are as calculated for a 1-m^3 isolator.

Further discussion of leak testing: the distributed leak test

The logical proposal has been made that every isolator should undergo a distributed leak test at OQ. Whilst the isolator may have reached the acceptance criteria for a specific leak rate measurement by, for instance, pressure decay testing, it may yet have holes of microbiologically significant proportions. Thus, the isolator should be further tested using a method that will detect individual leaks and hopefully qualify them as being not significant. This distributed leak test could be carried out using the helium method, or possibly DOP smoke. It would then provide support for subsequent in-service pressure decay tests.

Table 6.1

Class of Isolator	3	2	1
Hourly Leak Rate	$<1 \times 10^{-2}\,h^{-1}$	$<2.5 \times 10^{-3}\,h^{-1}$	$<5 \times 10^{-4}\,h^{-1}$
Percentage Volume Loss per Hour	$<1.0\%\,h^{-1}$	$<0.25\%\,h^{-1}$	$<0.05\%\,h^{-1}$
Volume Loss per Second	$2.8 \times 10^{-6}\,m^3\,s^{-1}$	$0.70 \times 10^{-6}\,m^3\,s^{-1}$	$0.14 \times 10^{-6}\,m^3\,s^{-1}$
Single–Hole Equivalent	464 microns	232 microns	103 microns

Further information and practical guidance on leak testing, including full formulae for calculations, are given in the *Yellow Guide*, third edition (Midcalf, Philips, Neiger, and Coles 2004).

Testing gloves and sleeves

The main structure of an isolator, even the flexible film type, is generally resistant to leakage, but the half-suits, sleeves, and, in particular, the gloves or gauntlets, are very much on the front line of battle. They are likely to be very close to the process within the isolator. They are continually stretched and flexed when in use, they may handle sharp tools and instruments, and they contain some potentially heavily contaminated items (i.e., the operator's hands). All of this makes the gloves and, to a lesser degree, the sleeves, the weakest link in the isolation chain and so deserving of special attention with regard to leakage. The *Yellow Guide* (Lee and Midcalf 1994) suggests that the gloves be tested every day, regardless of the type of operation being carried out, which is not an unreasonable stance. However, if we are to accede to this demand, then some very practical equipment will be required by the operators.

Most isolator manufacturers now produce devices for leak testing the sleeve and glove assembly *in situ*. One of the oldest methods has been to place a flexible diaphragm over the shoulder ring of an operating isolator and observe its behaviour. Any leakage of the sleeve in either a positive or negative isolator will cause the diaphragm to curve. The degree of deflection over a given period of time indicates the rate of leakage. There is now a generation of sleeve/glove test systems that fit into the shoulder rings of glove isolators. The complete sleeve is then inflated by external supply, sealed off, and the rate of pressure decay measured. This test can be carried out with the isolator running or static, but it has been suggested that the sleeve should be inflated into the isolator only in negative-pressure applications and that it should be inflated out from the isolator in sterile applications. The logic here is that if the sleeve does leak, then the flow will be in the correct direction to maintain the required containment. This would require the test plate to be used from within a sterile isolator; however, if leakage is found, then the glove or sleeve will be changed and the isolator resanitised in any event.

Devices that pressure test just the glove on an automatic cycle are in development, although a glove test system based on immersion of the glove in a nitrogen atmosphere and assessing leakage with an oxygen meter is now available. Whatever test method is adopted, the choice of gloves will always remain a balance between dexterity and robustness. There is no substitute for good operating procedures.

Further leakage considerations

The logic behind the operation of an isolator at a pressure above or below the surrounding atmosphere is that, should there be any leakage, it will always be in the correct direction. The product will be protected in the sterile, positive isolator, and the operator will be protected in the toxic, negative isolator. However, this is not necessarily the case, due to a phenomenon known as induction leakage. This may occur in two forms: (1) at the site of undetected holes and (2) at the site of known apertures, such as product output mouseholes (see Chapter 3, "Dynamic Mousehole"). The mousehole should be engineered to take into account induction leakage, but, in the case of an inadvertent hole, there could be a problem. If the movement of air close to the wall of an isolator has sufficiently high velocity, then its dynamic pressure might reduce and, in extreme cases, nullify the differential pressure across the isolator wall. If this occurs at the site of a hole, then there is potential for contamination to move against the static pressure differential.

Calculation using Bernoulli's law shows that leakage may occur when the isolator differential pressure is less than 20 Pa and the velocity of air movement across the face of the isolator is more than 6 m/sec. This is a fairly extreme situation, but it should be borne in mind that the movement of half-suits or sleeves may reduce the isolator differential at times, and possibly bring it close to the induction leakage situation. Just how significant induction leakage can be does not seem to have been quantified by any workers at this time, but it is the kind of factor to be borne in mind when considering isolator design and function.

When should leak testing be carried out?

A useful table suggesting when leak testing should be carried out in the lifecycle of an isolator is given in the *Yellow Guide*, third edition (Midcalf, Philips, Neiger, and Coles 2004), and reproduced herewith as Table 6.2.

Instrument calibration

If any instrumentation is fitted to an isolator, then it should be possible to calibrate that instrument to a known standard, not only following construction, but also at the recommended service interval. This is fairly simple with isolator pressure measurement: simply connect a calibrated micromanometer of suitable range in parallel with the isolator pressure meter, and plot any variation between the two across a suitable range of pressures. In some cases, the isolator instrument can simply be calibrated to read the same as the test instrument; in others, a calibration correction curve may be drawn. The same methods will probably be applied to pressure meters fitted across the HEPA filters on some isolators.

Airflow rate is not so easy to calibrate because the flow rates and the pressures involved are generally quite low; thus, devices such as pitot tubes

Table 6.2 When Should Leak Testing Be Carried Out?

Test	FAT	SAT/OQ	Routine Maintenance (e.g., 6-monthly)	General Routine In–Use Testing
Distributed Leak Test	Yes	Yes	N/A	N/A
Isolator Leak Rate Measurement	Yes	Yes	Yes	Positive isolator monthly, aseptic negative isolator weekly
Glove/Sleeve Leak Rate Measurement	Yes	Yes	Yes	Every session
Half-Suit Leak Rate Measurement	Yes	Yes	Yes	As for isolator
Leak Detection	Use to locate leaks if leak rate in excess of acceptance criterion			

are not often used. Rotating vane anemometers are sometimes used, but are inherently inaccurate in this situation. Hot-wire anemometers may be used, but they are very sensitive to position in the airflow. Enclosed turbine-type meters may be suitable in some situations. Another method commonly employed is the standard orifice plate. The pressure differential across a suitable orifice plate can be mathematically related to the mass flow. This system has the advantage of mechanical simplicity and robustness, combined with reduced interference from the effects of turbulence in the flow. The micromanometer used to measure the orifice differential can, of course, be used to calibrate the isolator pressure meter.

Any other instruments, such as temperature and humidity meters, should be calibrated by the means described by the manufacturers and the results reported.

Some users will go to considerable lengths to ensure accurate calibration of isolator instruments, and, whilst this may be commendable, absolute accuracy is not generally the raison d'être of such instrumentation. The purpose of monitoring the isolator pressure and flow is to reveal any changes from the validated norm, be they transient or trends. If the values are outside alarm or action limits, operators must step in immediately. If certain trends begin to show, these may indicate that a problem, such as a blocked HEPA filter, is developing and will require future action.

Airflow patterns

The path taken by the internal airflow of an isolator is significant. Clearly, in unidirectional downflow units, we would wish to demonstrate that laminarity is maintained through the isolator to a certain acceptable height above the base tray or process equipment. This is most simply done with smoke pencils, placed in various locations, while video film is taken to provide a validation record. Another method might be to survey the isolator volume using a Doppler shift velocity meter (e.g., Airflow Developments Ltd.) to map the airflow pattern, and possibly to use the data to produce a computer model. Indeed, in some cases, mathematical flow modelling might be accepted for OQ validation. Rotating vane and hot-wire anemometers may not be sufficiently sensitive for flow mapping in most isolators.

In turbulent isolators, we would want to check that there is uniform air distribution in the isolator, with no stagnant areas or standing vortices where contamination might build up. Again, smoke may be used to visualise the flow pattern, with DOP introduced in the inlet air duct, downstream of the inlet HEPA filter. Both buildup of smoke and the subsequent purge-out might be video recorded as validation data.

Miscellaneous checks

Particle counting

Since freedom from airborne particulate material is one of the principal functions of most isolators, it will usually be necessary to validate this parameter. This may be done by introducing the sampling head of an instrument, such as the Hiac/Royco, and counting at a number of sites within the body of the isolator. Ideally, this should be done both at rest and during operation to generate baseline data from which to establish the routine monitoring procedures, although measurement in operation would probably be included under the PQ. Most users record particle sizes at 0.5 micron and 5 micron and sample one cubic foot at between one and five sites within the isolator, depending on its size. Sampling takes place at working height, i.e., on the centerline of the sleeves, and should be isokinetic in unidirectional downflow isolators. However, recent thinking in relation to Class 5 of the new ISO EN 14644 standard (the old Class 100 under Federal Standard 209E) suggests that a sufficient sample volume must be taken to verify the level of 5 micron particles as being less than 29 per cubic metre. In practice, this means taking a sample of 1 cubic metre, which takes over half an hour with standard particle counters.

On a negative-pressure isolator, the counter may have to have an additional sample point to provide a return flow to the cabinet to prevent excessive differential pressure and possible pump failure (this is more relevant when sampling 1 m^3 for 35 minutes to comply with new GMP directives).

A further useful check on the isolator operation is particle recovery rate. This involves assessing the time taken for the isolator to clean down, usually from ISO EN 14644 Class 7 to Class 5. A small amount of DOP smoke is introduced into the isolator to raise the particle burden to Class 7, and then a particle counter is used to monitor the reduction in particles over time. The average isolator will clean down in a few minutes, while a unidirectional downflow isolator should take only a few seconds. This data will provide a useful performance comparison for subsequent routine maintenance.

CIP drainage and drying

In more complex isolators fitted with CIP/SIP systems, it may be necessary to check the action of CIP by a suitable challenge, as discussed in Chapter 8 on the PQ. The subsequent drainage action, however, is arguably part of the OQ. It can be checked visually, as can the drying process. The plan here would be to establish the worst points for drainage and drying, and then examine these visually after each process cycle.

Alarms

Most isolators will carry alarm systems with varying degrees of complexity, depending on the function of the unit. Some protocol will be required to check the operation of these alarms under all possible modes of failure.

Breach velocity

In containment isolators, the URS may stipulate a certain breach velocity — the velocity of air inward through a breach in the wall of the isolator. This breach is normally defined as the loss of a glove from the sleeve system, thus opening an aperture of about 100-mm diameter. The conventional requirement for minimum breach velocity is 0.70 m/sec, derived from BS 5726, the standard for biological safety cabinets. This is easily checked by removing a glove and measuring the velocity of the resulting airflow, perhaps one-third of the way across the diameter of the cuff ring, to gain an average value across the flow profile.

Interlocks

If interlocks are fitted to transfer devices, such as lockchambers, then these should be checked for potential failure mode. Where moving machinery is present in the isolator, there may be interlock devices, such as light beam arrays, across the glove ports to prevent entry. These, too, should be tested to find out if it is possible to defeat the system by simple methods.

Any other relevant parameters

Clearly, if the isolator is designed for a particular function, such as freedom from oxygen or water vapour, then this condition, along with any alarms, should be checked as part of the OQ, with subsequent in-process checks forming part of the PQ.

Manuals

Operating and maintenance manuals are required for any piece of equipment, from a simple incubator to a complex isolator, but the manuals for a product-licensed isolator system will form a part of the required documentation package. Historically, manuals have either been too generic to be useful, or have been submitted long after the equipment has been delivered. For these reasons, the subject of manuals should be addressed at the specification stage, and, ideally, a draft manual should accompany the DQ documents. The manual should certainly be available for operator training, starting at the predelivery inspection phase, as discussed in Chapter 8.

A typical manual might be structured as follows:

- Introduction, explaining in simple terms the purpose of the system and the main components and significant features. This might include a copy of the client specification, the manufacturer's quotation, and the client order to the manufacturer. A general arrangement drawing is useful in this section, as is a data sheet giving basic performance data and also contact information for servicing.
- A description of the general methods of construction and the materials used.
- A description of the ventilation system, with appropriate, as-built P and I diagrams.
- A description of the control and instrumentation system.
- Operational instructions for each part of the system — sleeves, suits, ports, etc.
- Planned maintenance schedule with daily, weekly, monthly, quarterly, and yearly checks.
- Description of maintenance tasks, such as leak testing and HEPA filter testing and changing.
- Spare parts lists with manufacturers' part numbers.
- Wiring diagrams and, ideally, circuit diagrams.
- P&IDs where appropriate.
- In-house and on-site commissioning data, including calibration certificates for the test instruments that were used.
- Information on any original equipment manufacturer (OEM) equipment that forms a part of the system, such as refrigerators, Millipore Steritest™ Integral pump, etc.
- Troubleshooting schedule.
- Log sheets for subsequent servicing, repairs, or changes.

Manuals should, of course, be clearly organised and written, and they should be securely bound between hard covers to preserve them for the lifetime of the isolator system.

chapter seven

Cleaning up: sterilisation and decontamination

This chapter is concerned with housekeeping in isolators — the regular physical actions required to maintain the internal environment of the isolator between batches of work. Where the process being carried out in the isolator is aseptic, then clearly the isolator must be sterilised before use; where the process is toxic, the isolator must be decontaminated before it can be opened for maintenance. Whatever the process, some form of general cleaning will probably be needed to prevent the buildup of contamination or product within the isolator.

Before discussing sterilisation, we should perhaps define the process more clearly, because the adjective *sterile* is an absolute, meaning the absence of viable organisms, whereas the methods used in isolators can never be regarded as absolute. We should perhaps correctly refer to these methods as sanitisation or disinfection, but the word sterilisation seems to be better accepted, provided that it is defined. The concepts of SAL and PNSU were mentioned in Chapter 1, but these refer specifically to the products, not to the isolator itself. The generally accepted scale for measuring the quality of a sterilisation process is derived from autoclave work, which is broadly analogous to isolator sterilisation.

The scale is given as the log reduction of the spores of a suitable resistant organism. Thus, we may place in the isolator a carrier that has 5 million spores of *Geobacillus stearothermophilus* or *Bacillus subtilis* on the surface. These particular organisms are chosen because they are nonpathogenic and very resistant to the typical sterilising agent. It is easy to quantify the number of spores on a carrier by drying a known volume of spore suspension, the number of spores in the suspension having been obtained by serial dilution. If, at the end of a sterilisation process, we find that the carrier now has 500 viable spores on the surface, we may say that a log 4 reduction has taken place. Carrying on with the autoclave analogy, if this log 4 reduction takes place in 20 minutes, we might say that the D-value is 5 minutes.

Since an SAL of 10^{-6} seems to be a generally accepted target for product sterility, then a log 6 reduction seems to be a realistic target for the isolator sterilisation process. The numbers are not directly related, of course, but they have a degree of equivalence. Thus, we may broadly define sterilisation for our purposes as log 6 reduction of *Bacillus subtilis* or *Bacillus stearothermophilus* spores.

In the case of decontamination for toxic isolator work, the process is not so easy to define in general terms. If the toxicity is due to a biohazard or pathogenic or recombinant organisms, then sterilisation, as defined above, will be required. If the toxic hazard is purely chemical, then the decontamination work should reduce this to a safe level, usually defined by standards such as the published OELs.

Sterilisation — manual wet processes

The actual sterilisation process will probably be chosen on the basis of speed, efficacy, and cost, with perhaps some consideration of safety, both local and environmental, and also compatibility with the structure of the isolator. Where time and cost are of the essence, then simple, manual wet processes will be chosen. There is a whole host of wet chemical agents to choose from, including the following:

- Formalin (40 percent formaldehyde solution with 10 percent methyl alcohol to inhibit polymerisation). This can be very effective, as it is sporicidal, unlike some wet agents, but it does need a high temperature and a high humidity to act with certainty. It is now considered to be toxic, possibly carcinogenic, and, certainly, it is very persistent in the environment. In fact, formaldehyde is banned in many countries and its use is likely to decline generally in the future. It is, though, an inexpensive agent, which will make it attractive in many cases.
- 70 percent ethyl alcohol (industrial methylated spirit or IMS). This is very cheap, it can be classed as nontoxic, and it has the major advantage of swift evaporation. However, the big drawback with alcohol is ineffectiveness. It will quickly kill the vegetative organism, but not the spores, in many cases. It is also flammable, but this aspect does not seem to create any real problems.
- 70 percent isopropyl alcohol (IPA). Much the same arguments apply to IPA, although it can be cheaper than IMS where duty (tax) is applied.
- Other aldehydes, such as glutaraldehyde (Cidex™). These are effective and medium priced, but, like formaldehyde, they are toxic and persistent, making them generally unpleasant to work with. Some operators have had serious respiratory afflictions caused by handling these substances.

- 3.5 percent peracetic acid (PAA) (Soproper ®). PAA, CH$_3$CO•OOH, is fairly cheap. It is very effective, being sporicidal, and breaks down quickly to relatively inert components — water, oxygen, and acetic acid (vinegar). However, it is quite corrosive and aggressive to handle.
- Sodium hypochlorite. This is quite cheap, fairly effective, but may attack some stainless steel surfaces.
- Chlorine dioxide (e.g., Virkon) is similar to sodium hypochlorite.
- Quaternary ammonia compounds.

Compounds such as these may be hand sprayed on the interior surfaces of the isolator or wiped onto all the surfaces with a lint-free cloth, taking care to cover occluded areas and surfaces, such as the seals on ports and doors. As with cleanroom maintenance, it is common to rotate several different agents to reduce the potential for the buildup of resistant organisms.

With a large isolator, hand spraying would be tedious and operator sensitive, but with small isolators, such as hospital pharmacy dispensing, hand spraying proves quite adequate. It should be noted that flexible film isolator canopies are somewhat susceptible to the effects of solvents, like alcohols, that may leach out the plasticiser and cause cracking or surface filming. In this case, the use of such solvents should be reduced to a minimum, but it is still possible to spray flexible canopies once a day without long-term damage. The ductwork and HEPA filters are not reached by manual methods and validation is probably not easy, particularly for larger isolators; thus, its use is generally limited to operations such as dispensing.

Fogging processes

Formaldehyde

Biological safety cabinets have for a long time been sterilised by the very simple process of boiling off a small volume of formaldehyde solution within the sealed cabinet. The cabinet is left static overnight, after which the fan is switched on to purge with air to a suitable exhaust duct. Such a process can also be used in isolators. The treatment is very effective in terms of biocidal activity, and the equipment is inexpensive, but doubts are growing over the general safety of formaldehyde, as mentioned earlier. There is also a tendency to form white deposits of paraformaldehyde in the isolator, and it can take a long time to purge down to low levels of formaldehyde.

PAA and other agents

At least one automatic fogging device is commercially available for treating isolators, the French SPRAM machine, available from La Calhène SA. This produces a fine mist of the sterilising agent directly into the isolator, using compressed air as the driving force. A very high concentration of the agent can be achieved very quickly and, as a result, very quick kill times can be achieved. The suggested sterilising agent is 3.5 percent PAA, although any other liquid agent can be used. The disadvantages of such methods include an increased probability of corrosion problems, especially with internal equipment, and possible difficulties with validation.

Fogging methods are no doubt more effective than manual ones, since the full volume of the isolator is treated; thus, all parts of the isolator body and its contents should be reached. However, sterilisation of the HEPA filters and any ductwork relies mainly on diffusion, and so there is some scope for failure.

Gassing systems

The sterilisation of isolators by gas-phase treatment has a number of attractions:

- The gas can be introduced ahead of the inlet HEPA filter and it can be removed after the exhaust HEPA filter; thus, the entire system, including the filters and their associated ductwork, can be treated. Figure 7.1 illustrates the connection of a gas generator to a simple turbulent flow isolator.
- If the agent is purely in the gas phase, its behaviour will be reasonably predictable in terms of buildup, dwell time, and purge-out. This makes the process easier to validate than other methods.
- Gas-phase agents are much less likely to cause corrosion of the isolator and, more particularly, of the equipment inside the isolator. Thus, electronically controlled equipment can be safely used in gas-sterilised isolators.
- Automatic gassing methods are likely to be safer for the operator, since little or no contact with the agent is required.
- At least some of the gassing systems available are also more likely to satisfy GMP requirements for documentation of sterilisation cycles.

Recent developments in gas-phase sterilisation

Much research and development work has taken place in gas-phase sterilisation, particularly using hydrogen peroxide vapour (HPV), in the last five years. There have been advances in the understanding of the process, there have been improvements to the equipment, and the regulatory authorities have placed more emphasis on the need for gas-phase sterilisation. Perhaps the most significant advance results from the work by the British company, BioQuell PLC, formerly known as MDH Ltd. This work indicates that the levels of HPV that have been reported in the past cannot, in fact, be present in the true vapour phase. Levels around 2500 ppm have been reported, but it has been shown that the saturation level of HPV under normal conditions can only be about 250 ppm. Where, then, is the hydrogen peroxide that has

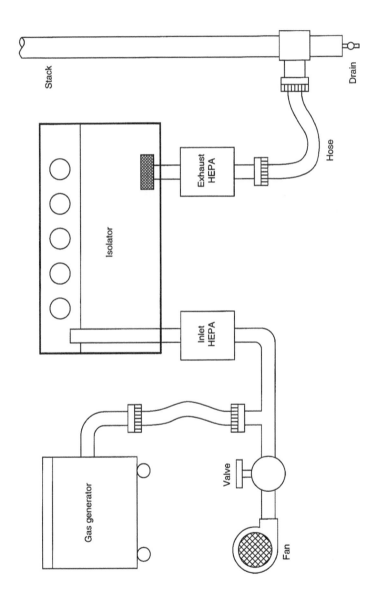

Figure 7.1 Hose Connections for a Turbulent Isolator. A schematic diagram of the hose connections for gas sterilising a turbulent flow isolator using an open-loop gas generator such as the Sterivap or Citomat.

been evaporated by the gas generator and delivered to the isolator? The answer appears to be that the vapour actually condenses on the isolator surfaces, but in the form of droplets, which are of the order of 1 micron across and thus not directly visible as condensation. This phenomenon has been termed *micro-condensation*. Furthermore, it has been shown that this condensation occurs preferentially on the surface of microorganisms, thus inactivating them.

This work is highly significant and shows that, although the HPV sterilisation process appears to be dry, it is in fact a wet process, at the microscopic scale. This being the case, the parameters of the gassing process can be optimised to induce fast micro-condensation on the isolator surfaces, and thus produce fast but reproducible gassing cycles. Unfortunately, the notion of micro-condensation has produced something of a schism between the manufacturers of gas generators, one party declaring that the process is dry whilst the other claims that it is wet. The debate is interesting, but, in the final analysis, the hydrogen peroxide sterilisation process can be validated perfectly well for both parties. It is the opinion of the author that the two supposedly different processes (wet versus dry) are in fact the same process, and any differences are more commercial than physical.

Another area of development, and one that may actually be more significant than the micro-condensation issue, has been in the area of BIs. Operators of gassing processes have reported concern over the survival of BIs used in the routine revalidation of well-established cycles. This leads to serious questions about the reproducibility of the gassing cycle; however, recent research shows that the fault may well lie with the BIs. Examination of a series of BIs has shown that whilst some may present an even monolayer of spores to the gas, many have multiple layers, significant clumping, and include much debris from the vegetative cells. Such BIs would give very variable results under gassing conditions. The PDA has set up a task force to report on this work and, at the time of writing, a draft paper has been issued for comment.

The subject of the practice and validation of gas-phase sterilisation has become lengthy to the extent that an entire book could be produced on the subject. Indeed, it seems likely that a monograph will be produced in the near future.

The following are accounts of some isolator gas sterilisation devices. The list is illustrative, rather than exhaustive.

La Calhène SA Sterivap®

Isolator gassing systems divide broadly into two groups: those with an open loop that exhaust the sterilising gas to the atmosphere, and those with a closed-loop cycle requiring no exhaust to the atmosphere. One of the earliest open-loop gas generators, the La Calhène SA Sterivap, was developed in

France, initially for the treatment of flexible film animal isolators with formaldehyde. Since then, it has been used successfully with 3.5 percent PAA for sterilising isolators of all types.

The process diagram (see Figure 7.2) for this gas generator is very simple: It consists of a glass vessel of about 2-L capacity, surrounded by a heating jacket that holds the contents of the vessel, 3.5 percent PAA solution, at 50°C. Compressed air is blown into the top of this vessel at a rate of about 2.5 m^3 per hour, where it picks up PAA vapour from the surface of the liquid. The vapour-laden air passes through two simple condenser columns that cool the vapour to ambient temperature and remove any resulting condensate. The dry vapour is then introduced into the isolator ahead of the inlet HEPA filter. It passes through all the ductwork to enter the body of the isolator, and then leaves the isolator system after the exhaust HEPA filter. Finally, the exhaust-sterilising gas is ducted to the atmosphere. The system has no real regulation of the relative humidity of the sterilising gas; thus, the gas may condense in the isolator if the ambient temperature is below the dewpoint for the vapour. The dewpoint of the gas will vary with the relative humidity of the compressed air used. Gas condensation will lead to a reduction of gas concentration, and eventually can lead to pooling of liquid that may not evaporate fully during the purge phase of the cycle.

To operate the machine, the gas hoses are connected to the isolator, the reservoir is filled with 3.5 percent PAA, and the compressed air supply is switched on. The in-built regulator valve and flow meter are used to set, manually, an airflow rate of 40 L/min, thus balancing the resistance of the filters and ductwork. The generator will continue to blow compressed air across the reservoir for a gassing time period set by the operator; then it will operate solenoid valves and blow only air through the system to purge the vapour from the isolator.

PAA vapour is a good sterilising agent and can achieve log 6 reduction in a matter of a few minutes. However, the Sterivap has a very low flow rate, and thus the time to build up the gas concentration to a lethal level, and the subsequent purge-out times, are relatively long. This effectively limits the machine to use in isolators up to a maximum of about 5-m^3 volume. The manufacturer suggests a cycle time of about five hours for a 3-m^3 (single half-suit) isolator.

The Sterivap has little in the way of instrumentation and performance monitoring, so repeat cycles are not easy to validate. The supply of compressed air must be pressure stabilised if the flow rate is to remain constant, and a visual instrument is the only indication of this important parameter. The operator is required to measure the volume of sterilant poured into the vessel before the cycle, drain the vessel as well as both condensers, and measure the amount remaining after the cycle. The volume lost is then assumed to have been vapourised into the isolator.

Despite these shortcomings and the very simple design, the Sterivap is a surprisingly effective device, compact in form and very inexpensive when compared with more recent, complex gas generators. It is available in two

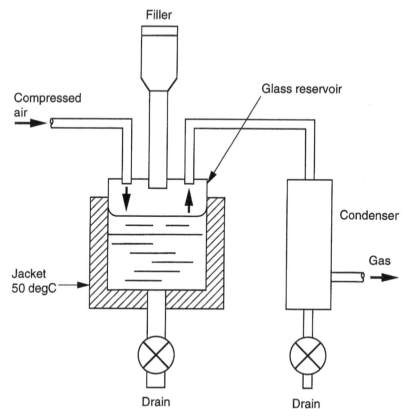

Figure 7.2 A Schematic Diagram Showing the Operation of the La Calhène SA Sterivap Gas Generator.

versions: one is bench mounted and the other, which has some further sophistication, is floor mounted. It has found use in numerous small isolator installations around the world.

Astec Microflow Citomat

At the time of writing, the Citomat gas generator is not available commercially, but a very similar generator is likely to be manufactured in the near future. This type of generator represents a very significant stepping stone between the very simple Sterivap and the more complex closed-loop generators that follow. For these reasons, a description of the Citomat is included.

The Citomat (Figure 7.3) was developed primarily as an advance on the Sterivap, to give a higher output rate, better control and monitoring of the operational parameters, and no requirement for a compressed air supply. The design is, nonetheless, quite simple and does not incorporate any form of air preconditioning; thus, it relies on the air in the isolator room being of

Figure 7.3 The Astec Microflow Citomat Gas Generator. Air enters the generator through the circular prefilter on the front face, the sterilising agent is placed in the beaker on the top of the enclosure, and gas leaves through the union coupling on the right-hand face. The touch control panel is seen on the angled top panel.

an acceptable relative humidity and temperature. Figure 7.4 is a schematic diagram of the Citomat system.

The heart of the Citomat is a double-walled chamber: the inner of stainless steel and the outer of rigid PVC. The inner chamber is wound with

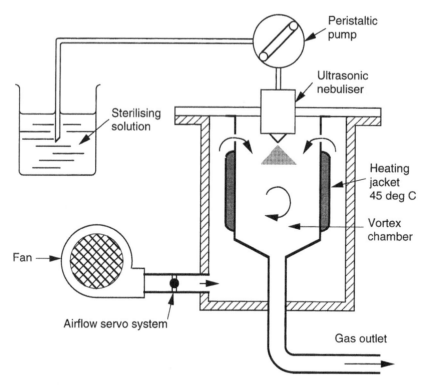

Figure 7.4 A Schematic Diagram Showing the Operation of the Astec Microflow Citomat Gas Generator.

heating tape, which is controlled to 45 ± 3°C. A centrifugal fan blows air into the space between the chambers, where the air is warmed before spilling into the top of the inner chamber. Two angled vanes impart a rotation to the air in the inner chamber. The airflow rate is servo controlled to 20 ± 2 m³/ h. At the top of the chamber is an ultrasonic nebuliser, which is fed with an aqueous sterilising solution by a peristaltic pump at the rate of 2 ± 0.2 ml/ min. The nebuliser breaks up the solution into a mist, which then evaporates completely in the warm air vortex. A hose carries gas to the isolator, as with the Sterivap, and then the exhaust gas is ducted to atmosphere.

If the laboratory air is at 20°C and 50 percent relative humidity, then the Citomat will add 5 g of aqueous solution to each 1 kg of air, and raise the relative humidity to 85 percent once the gas has cooled back to 20°C in the isolator. This is shown in the following psychrometric chart (Figure 7.5).

Thus, the generator will work satisfactorily within a certain envelope of relative humidity and temperature, but outside this, condensation will occur in the isolator. The concentration of the gas delivered is about 350 ppm by weight.

Originally, the Citomat was designed primarily to use 3.5 percent PAA as the sterilising agent, but a subsequent search for a less corrosive and aggressive agent resulted in the production of a mixed agent, branded

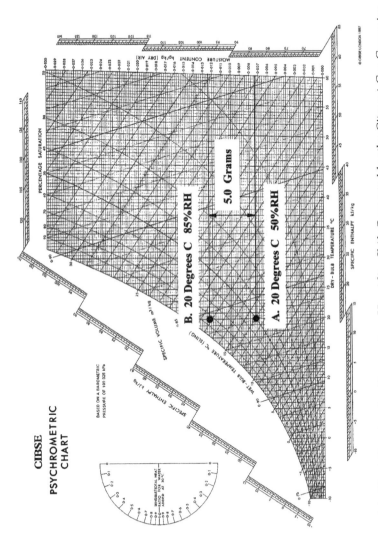

Figure 7.5 A Psychrometric Chart Showing the Temperature-Humidity Shift Generated by the Citomat Gas Generator. Laboratory air at 20°C and 50 percent relative humidity is loaded with 5 g of aqueous solution per kg of air, moving the resulting vapour stream to 20°C and 85 percent relative humidity once it has cooled to ambient in the isolator. This value is thus 2.5°C above the dewpoint. (Courtesy of the Chartered Institution of Building Services Engineers, London.)

Citanox® by the makers. This consists simply of a solution containing 10 percent hydrogen peroxide and 1 percent PAA by weight. It seems to be almost as swift in action as PAA, but less corrosive; indeed, in the vapour phase, it appears to be quite compatible with the sterilisation of electronic equipment. The gas breaks down fairly quickly, depending on the nature of the surfaces that it encounters, to water vapour, oxygen, and a trace amount of acetic acid; thus, it is environmentally acceptable.

The gassing process with the Sterivap and Citomat consists of three phases: the buildup of gas concentration, a dwell time for the sterilising action to take place, and, finally, a purge-out period to remove the gas prior to starting work in the isolator. The time taken for the buildup depends on the flow rate of the gas generator, the volume of the isolator, and the breakdown rate of the gas. Of these, the first two can be obtained easily, while the third is dependent on many factors and is not easy to measure directly. Infrared absorption is probably the best direct method available, but simply measuring the relative humidity whilst using pure water on a gassing cycle at least provides some indication of the gas behaviour. The change in relative humidity produced by the Citomat feeding to a 0.50 m³ two-glove isolator is shown in Figure 7.6. This shows the classic buildup curve from 30 to 65 percent relative humidity after 15 minutes, followed by an asymptotic decline during the purge phase. Direct measurement of the gas concentration at the same time would probably have produced a similar curve, but reaching to lower absolute concentration due to gas breakdown and a longer purge curve due to adsorption and subsequent degassing in the isolator.

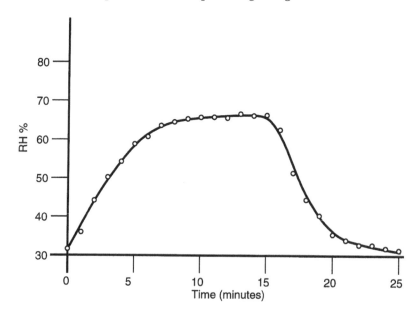

Figure 7.6 The Typical Relative Humidity Curve Given by a Citomat Gas Generator in a 0.5 m³ Isolator.

The microbiological effect is shown by the kill curve in Figure 7.7, which was produced with a Citomat and Citanox in a 0.50-m^3 flexible film, two-glove isolator. This curve is obtained by exposing spore strips, in this case *Bacillus subtilis*, to varying times of gassing and then counting the survivor rate. The resulting curve shows the dynamics of the kill and is a better indicator for validating suitable gassing cycles than simple growth/ no growth tests. Using the kill curve, we can establish what time period is needed for log 6 reduction to take place, and a suitable gassing-time safety margin, for instance, a further 50 percent, may be added for the production SOP.

Two curves are shown in Figure 7.7: one for spores on aluminium carriers, the other for spores on paper carriers — the latter being slightly harder to kill, partly because the spores will be deep within the fibres of the paper and partly because the cellulose tends to deactivate peroxides. The manufacturer suggests a total cycle time of around 2.5 hours in a 3-m^3 (single half-suit) isolator.

A significant feature of the Citomat is the chart recorder that gives a hard copy record of the operating parameters for each cycle. The multipen recorder charts the chamber temperature, airflow rate, liquid flow rate, and

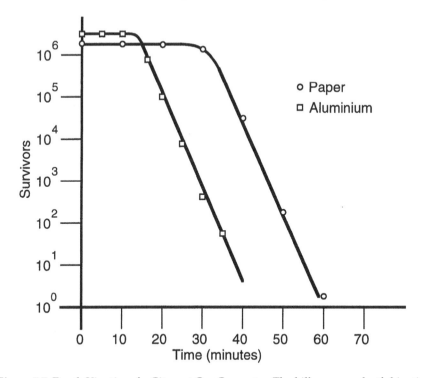

Figure 7.7 Death Kinetics of a Citomat Gas Generator. The kill curve or death kinetics given by a Citomat gas generator using Citanox as the sterilising agent and *Bacillus subtilis* spores as the BI in a 0.5-m^3 isolator. Aluminium spore carriers seem easier to kill in this case.

gassing/purging periods. This information can then be attached to the product batch release documents as part of continuous QA. Alternatively, the RS232 port can be used with a suitable emulator and the data downloaded to a PC.

Steris VHP 1000™

At about the same time that the Citomat was under development, AMSCO (now Steris) was working on a rather larger, and much more sophisticated, closed-loop, hydrogen peroxide gassing system that was named Vapor Hydrogen Peroxide (VHP) 1000. Figure 7.8 shows a VHP 1000 together with a flexible film isolator. This unit introduces a new step in the gassing cycle, by first drying the air in the isolator system using a chemical bed. This allows a much greater concentration of vapour to be carried in the air without condensation developing, and makes the initial relative humidity and temperature of the air in the isolator practically irrelevant.

Hydrogen peroxide is flash evaporated into the dry airstream from 30 percent (100 volume) solution. The resulting gas is passed around the isolator to be treated and then returned to the generator, which contains a catalyst bed. This breaks down the gas to water vapour and oxygen to complete the cycle, and forms a closed-loop system with no environmental emission. The hydrogen peroxide solution is supplied by Steris in special, sealed cartridges that fit into a compartment on the front of the machine, so that the operator does not have to handle the material at any stage. The control system

Figure 7.8 The Steris VHP 1000 Hydrogen Peroxide Gas Generator in Use with a Flexible Film Isolator. (Courtesy of Envair Ltd.)

provides a digital readout of the cycle parameters, which can be downloaded to a PC, and emphasis is placed on the ease with which the system can be validated to FDA requirements. The manufacturer says that the VHP 1000 can handle isolator volumes up to 30 m³, which is very large in isolator terms. As a result of these advanced features, the VHP 1000 has sold well and is to be found in pharmaceutical plants throughout the world, not only for use in isolator sterilisation but in other applications as well.

BioQuell Clarus™

The development of the BioQuell Clarus was the result of an investigation into why the industry was experiencing problems on revalidating previously qualified hydrogen gassing cycles. In the first instance, the focus on the investigation was on quality issues in resistance variation of BIs. However, the investigation extended to the process itself with studies to establish the optimum conditions for biological decontamination efficacy. BioQuell noted that kill time at 6 log reduction of *Geobacillus stearothermophilus* and repeatability of kill time was faster and more consistent when conditions reached saturated equilibrium vapour pressure; hence, micro-condensation was formed. It became evident that a different set of process conditions was required, which needed a different system design to meet the requirement.

Although there was evident success of the Steris VHP 1000, this system was a single closed-loop system that was not designed for, nor could control, conditions of micro-condensation. This has led BioQuell to develop a closed-loop, HPV generator that incorporates a number of advances that primarily maintain and control micro-condensation conditions in the bio-decontamination cycle phase.

- The Clarus is based on a dual-loop, closed-circuit design with two changeover air/gas circulation routes for different phases of the process, so that each circuit can be optimised for the process requirement at that stage.
- During the initial cycle phase of relative humidity and temperature conditioning, the air circulation path is through a refrigeration process dehumidifier, which is in continuous operation. In contrast, the VHP 1000 has a chemical dryer, which requires a regeneration period after ten hours or so of operation and requires 18-hour regeneration time. This is a major disadvantage where continuous production is taking place.
- Once the preconditions have been achieved, the gas circulation path changes over with a vapouriser in circuit. At this time, there is no catalyst in the same circuit, so vapourised hydrogen peroxide is permitted to build in concentration to saturated conditions, resulting in micro-condensation (typically nonvisible) developing on the surface of the separative enclosure requiring bio-decontamination.

Effectively, the Clarus uses the condensation process to deliver the hydrogen peroxide disinfection agent directly onto target surfaces.

- Following the bio-decontamination phase gas residuals are removed by changing the gas circulation route back to the conditioning circuit, which includes a catalyst. This cycle phase, (d), is called aeration.

- Aeration is a process of reevaporation of micro-condensation from surfaces and breakdown of hydrogen peroxide to water and oxygen by a catalyst. The generated water from the aeration process is removed by the conditioning circuit refrigeration dryer and collected in a waste bottle.

- Clarus does not continuously break down and replenish the HPV in the system. Instead, the vapour only passes the catalyser at the end of the cycle, the gas being topped up by the evaporator as need be, during the gassing phase. This results in a major reduction in the consumption of hydrogen peroxide solution.

- The BioQuell Clarus range of gas generators includes two units that are for supply to clients, Clarus L and Clarus C, and two units that support a gassing service, Clarus R and Clarus S, provided by Bio-Quell and authorised parties.

- Clarus L is typically used for bio-decontamination of laboratory equipment, safety cabinets, incubators, ventilated cage racks, small isolators, etc. Clarus C is used for isolators requiring a high level of parametric control and, because of the evaporation and control, capacity can be used for cleanrooms up to 200 cubic metres. Clarus R is theoretically infinitely scalable in capacity and has been validated up to 750 cubic metres cleanroom volumes. Clarus S is primarily for biological safety cabinets.

- The Clarus C uses easily obtainable industry-standard bottles of 30 percent hydrogen peroxide solution rather than specially developed, dedicated cartridges. Clarus L uses a dedicated fill volume charge for sealed handling and is without residuals, as the 10- to 175-ml range of bottles provide the optimum amount to be used for each developed gassing cycle; a bar code reader assists traceabilty.

- All Clarus hydrogen peroxide gas generators include an integrated electrochemical cell H_2O_2 gas monitor that gives a direct readout of the gas concentration of the HPV, typically measured directly in the separative enclosure. H_2O_2 gas concentrations are printed together with all other critical cycle references and real-time data.

- The Clarus C offers an optional condensation meter which, when fitted inside the isolator, can control and optimise the delivery of hydrogen peroxide to the system, holding the concentration at the point of micro-condensation.

- The Clarus range is supported by gas distribution devices utilising rotating jet nozzles to provide very efficient gas distribution without the cleaning challenges provided by the traditional gas mixing stirrer fans mounted in the separative enclosure.

- Clarus gas generators are configured to operate with aeration assistance devices to reduce the overall cycle time. Assisted aeration may be via the enclosure HVAC or via catalyst systems that work in association with the gas generator catalyst.

Figure 7.9 shows the overall appearance of the Clarus C gas generator. The Clarus gas generator range utilises Siemens PLC control, with digital readout and alarm of the operating parameters and automatic printout. Comms ports, RS232 and RS 485, are provided for connection to remote equipment, data logging and interface to other PCs.

The bio-decontamination cycle can be configured to run at positive or negative pressure. Clarus C has circulation flow rates of 30 m^3/h and variable H$_2$O$_2$ injection rates up to 8 g/min. Clarus L has circulation flow rates of 20 m^3/h and a fixed H$_2$O$_2$ injection rate at 2.8 g/min.

Clarus C holds up to 99 cycles in a software HMI library and Clarus L holds 10 cycles. Each cycle has a defined recipe with input parameters developed through a process of cycle development before PQ qualification.

Calculations for injection rates to achieve micro-condensation conditions for given enclosure volume are completed using a unique computer program. The science and basis of the program follows the monograph by Watkins et al. (2002).

HPV seems likely to become fairly standard as the preferred method for bio-decontaminating industrial-scale isolator systems, and so it is worth examining the process cycle of Clarus C in a little more detail at this stage. Figure 7.10 is the schematic flow diagram.

- The system is driven by a multistage centrifugal fan capable of developing up to 3 kPa to overcome the pressure drop through the Clarus itself, the isolator and its HEPA filters, and all of the associated ductwork.
- Through the conditioning circuit, the fan delivers air to the catalyser and then to a dehumidification column with a refrigerant dryer, which cools the air to 10°C. Any condensate is pumped to the waste collection receiver bottle. In addition, any residual hydrogen peroxide remaining in the Clarus C dosing column at the end of the cycle is pumped to the waste collection bottle so that all new cycles start with fresh valid agent.
- Air then passes over a reheater to bring back the delivered air temperature to normal conditions before delivery to the enclosure.
- Airflow rate is measured by passing return airflow through a standard orifice plate and monitoring the pressure drop across the plate with a high-resolution pressure transducer.
- An integrated pressure control system takes air in via a HEPA filter and expels an air/gas mixture out via a catalyst/HEPA filter to maintain set pressures (positive or negative) whilst in closed-loop

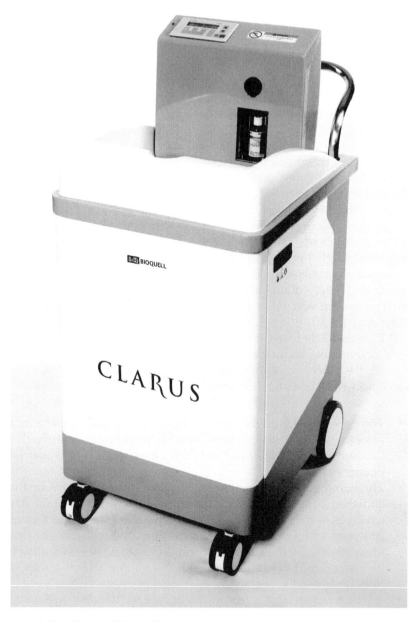

Figure 7.9 The BioQuell PLC Clarus Hydrogen Peroxide Gas Generator. The compartment in the lower right front panel carries the supply of 30 percent peroxide solution. The smaller compartment on the left carries the condensate collector. The gas supply and return hose connections are out of sight on the back of the enclosure in this view. (Courtesy of BioQuell PLC.)

Schematic Flow Diagram

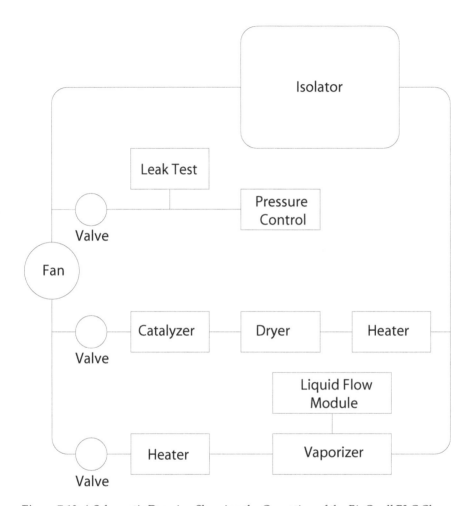

Figure 7.10 A Schematic Drawing Showing the Operation of the BioQuell PLC Clarus Hydrogen Peroxide Gas Generator. (Diagram provided by BioQuell.)

operation. A reference tube is connected between the gas generator and enclosure to facilitate pressure monitoring and control via the Clarus system.
* The gas generator will drive air around this circuit until it has been dried down to the required humidity, typically 40 percent RH. The catalyser and dryer are then closed off and the flow path changes to the vapouriser circuit.

- The dry, warm air now enters the evaporation module (vapouriser), where hydrogen peroxide solution is delivered to a hot plate and flash evaporates at 125°C. The final gas generator delivery temperature is 60°C.
- The hydrogen peroxide is transferred from a standard bottle reservoir via a peristaltic pump to a dosing column that uses pressure transducer measurement, operating under the command of the PLC controller, which calculates and accurately delivers at defined injection rates. The system constantly accumulates the delivery rate and modulates to ensure accurate total delivery weight (grams) for the complete cycle.
- The HPV then passes through the isolator via heat-traced hoses, to limit any losses via condensation on route, and is returned to the Clarus once more.
- A HEPA filter on the gas return prevents the ingress of particulate matter that might poison the catalyser and contaminate the vapouriser.
- At the end of the gassing phases, the gas path switches back to the conditioning circuit, initiating the aeration phase with the air/gas mixture passing over the rare-metal catalyst bed, which breaks it down to water vapour and oxygen.
- All cycle phases can be either controlled to PLC set time points or parametrically to conditions measured by critical/calibrated instruments.
- As with Steris for the VHP 1000, the Clarus manufacturer, BioQuell, is able to advise clients on the application and the validation of the process. Full support documentation (IQ, OQ, O and M Manual) is normally provided, as well as QA and operator training.

HPV as a sterilising agent

As a sterilising agent, HPV has considerable benefits. It is highly sporicidal even at very low concentrations; at high concentrations, it can give very fast log reductions. Whilst the 8-hour OEL is given as 1 ppm, it is relatively nontoxic to humans, and it is also relatively noncorrosive to materials. A major benefit in these days of environmental concern is the fact that it is effectively unstable and breaks down more or less quickly to water and oxygen. It is not, however, a panacea to all the problems and issues of isolator sterilisation, and users should be aware of the following considerations:

- HPV is fully stable, in bulk, up to 300°C (Schumb et al. 1956). However, the breakdown to water and oxygen is catalysed by a wide variety of material surfaces, especially heavy metals and organics. The speed of this reaction is governed not only by the composition of the surface, but also by its nature. Thus, the surface finish of, for instance, a sheet of stainless steel may produce a varying breakdown of the HPV, depending on the degree of polish.

- HPV has poor penetrative properties, in terms of its ability to reach into recesses and blind holes. This may be linked to instability, the gas condensing in areas of low circulation. Thus, quite powerful stirring is recommended in larger isolators, and those loaded with equipment and materials, in order to circulate the gas freely over all surfaces. Such stirring may be provided by internal fans, which add kinetic energy to the gas in the isolator. Special distributive nozzles have also been developed to blow the gas vigorously around the isolator. These are motorized in two planes and thus disperse the gas widely to produce an even BI kill throughout the isolator and its load.
- In some cases, direct injection of the gas through the wall of the isolator is used, in order to get the gas concentration raised quickly. The gas exits via the isolator exhaust in the normal way, and diffusion is found to take the gas across the inlet filters and kill BIs placed in the inlet plenum chamber.
- HPV is generally noncorrosive, but it does affect natural rubber latex and nylon plastic in particular. Latex rubber gloves can be used with hydrogen peroxide gassing, but only for a limited number of cycles.
- Several materials that may be found in isolators can cause rapid breakdown of the gas with apparent partial inactivation of the sterilisation process. These include cellulose materials (paper packaging) and copper and its alloys (some electrical equipment). Organic materials, particularly protein and those materials containing catalase, accelerate breakdown, and gross organic contamination will tend to render the sterilisation less active.
- Long delivery hoses can apparently cause a considerable reduction in the concentration of the gas delivered to the isolator. Trace heating of long pipe runs has helped to alleviate this problem.
- Finely divided titanium dioxide catalyses the breakdown of hydrogen peroxide very well. There have been cases where titanium dioxide was used to test HEPA filters, which then passed no hydrogen peroxide whatsoever, to the puzzlement of the operators.

All of these factors need to be taken into consideration when designing and validating an isolator system for hydrogen peroxide sterilisation.

Other gas-phase sterilising agents

Steam/HPV

Some workers have reported very fast sterilisation times using a mixture of HPV and atmospheric steam (Pflug et al. 1995). They indicate that this method can achieve concentrations of hydrogen peroxide as high as 7500 ppm (0.75 percent). At this level, they achieve a D value (1 log reduction) for *Geobacillus stearothermophilus* of only 12 seconds. This results in a log 12 reduction time of a mere 2.4 minutes; when a concentration of 2500 ppm is used, the log 12 reduction time is still only 5.7 minutes. These times do not include any preconditioning, concentration buildup, or subsequent purging times, which must be taken into account when judging the merits of such a method. The potential for condensation and corrosion must be significant using these conditions and this, too, must be evaluated.

Ozone

Ozone appears to have great potential as a gas-phase sterilant for use in isolators, but reports vary widely regarding both its efficacy and its toxicity. Some workers report very quick microbiological kill, others very slow. Some workers suggest that ozone has low human toxicity; others give it very high human toxicity. Clearly, much work remains to be done on ozone before it can be evaluated as a sterilising agent.

Ultraviolet light

Although not a gassing method as such, the use of ultraviolet light as a sterilising agent has been discussed in relation to isolation technology. Davenport and Melgaard (1995) reported on the use of ultraviolet light for the transfer of presterilised packages, such as BD Hypak™ syringe tubes, into isolators. The object of the process was not a bulk sterilisation but the treatment of the pack surface, which is exposed briefly to a nonsterile atmosphere during manual transfer.

Exposure to ultraviolet light in the wavelength range 200 to 300 nm, at an intensity of around 4000 mW/cm^2, and using *Bacillus subtilis* spores as the test organism, gave a log reduction of about 5 in 60 seconds. This represents an effective sterilising process, but the method does have limitations.

Clearly, the irradiator must be designed to illuminate the packages from all sides evenly, bearing in mind the effects of the inverse square law on radiation intensity. Other considerations, such as humidity, temperature, and packaging material, have an effect on the process and must be taken into account. However, ultraviolet irradiation evidently has a useful place in some areas of isolation technology and should be considered as a potentially valuable sterilisation method.

Validating the sterilisation process

Much could be written about the validation of the gas-sterilisation process, and so the following is a relatively brief summary of the considerations. The basic function of the validation exercise is to develop validated gassing cycles. The central element of the cycle development process is to find the time taken to achieve a given log reduction of a chosen BI — the kill time. This is done by exposing BIs to the gas for various time periods to establish at what point the required log reduction takes place, a process that has been termed the *sublethal study* or the *partial cycle study.* Figure 7.7 shows the results of a sublethal study, plotted out to form a typical kill curve or death kinetics curve.

A safety margin is then normally added to the kill time, and the resulting time is then used in subsequent operation. Cycle development should be undertaken with the isolator in the normal operating condition with respect to loading with equipment or materials. The loading pattern used in validation must be carefully documented and then replicated for subsequent in-service gassing cycles. Having developed a given cycle, probably under the OQ protocol, operators may then run three consecutive cycles to further verify the process under the PQ protocol. It is worth noting that the PIC/S document suggests that having established a kill time, the three PQ cycles should be run at this time. Any safety margin added to this time for the operation is at the discretion of the QA department — the suggested margin is between 50 and 100 percent.

Prior to cycle development using BIs, a number of prestudies are usually carried out to establish where gas circulation may be poor and thus to provide worst-case locations for BIs. Such studies include:

- Visualisation of the gas flow in the isolator, using DOP smoke or water mist to illustrate the gas flow pattern.
- Thermal mapping using thermocouple arrays or thermal loggers, to establish any hot spots where micro-condensation would be less likely to take place.
- Gas distribution test using chemical indicators, to show up any areas where gas penetration is poor.

The number of BIs that may be used for each test cycle will vary with the size of the isolator and the nature of the load, but, as a rough guide, users will normally place about 25 BIs in a four-glove isolator (1 m^3) and perhaps twice that number in a double half-suit isolator (5 m^3).

Needless to say, all of the gassing cycle operating parameters, load configuration, BI information, date, time, operator, etc., must be fully documented. Modern gas generators will automatically log much of this information on a cycle printout sheet.

CIP in isolators

Where isolators are to be used for industrial-scale operations, such as the aseptic filling of vials, the sterilisation process must be preceded by a cleaning process. In order to minimise manual intervention and to maximise the ease of validation, CIP is becoming the norm. If CIP is to be incorporated, the isolator must be designed from the outset with this factor in mind and incorporated in the isolator P&ID. This topic has been well presented by several workers, notably Melgaard (1995), who suggests that there are two aspects to isolator CIP. The first is the reduction of the bioburden to a minimum prior to gas-phase sterilisation (SIP). The second is removal of product residues to prevent cross-contamination following product changeover. A further aspect where toxic products are being handled is decontamination-in-place (see below).

CIP design starts with the shell of the isolator, which will need generous edge and corner radii to reduce potential water traps as far as possible: 25-mm radius seems normal, but 50-mm radius is sometimes required. The floor of the isolator will need careful consideration so that the washwater flows easily to a single drain-down point. The suggested floor slope is 4 percent (1 in 25, or 2°). To avoid another potential trap, the drain point should be fitted with a flush-surface sanitary valve, such as that made by ASEPCO. This valve may need some form of guard to prevent shards of broken glass entering the waste system and damaging the pumps. The drainage system will need to cope with high volumes of liquid, of the order of hundreds of litres per minute, and should be sized accordingly.

Clearly, ledges and shelves should be avoided in the structure and, if unavoidable, they should slope toward the drain point. The design of window fittings and seals, particularly those which must be openable, should avoid ledges and crevices; the same is true of doors and transfer ports. The equipment inside the isolator should also be designed without ledges and shelves, for free drainage through the structure.

The siting and design of the spray devices used will need careful consideration, with emphasis on the more critical parts of the assembly, such as filling heads. A wide variety of spray balls, spray bars, and rotating sprays are available, and the advice of a specialist in CIP work may be worth the investment to optimise design here.

HEPA filters must, of course, be protected from wetting by the CIP system, and various methods are used, incorporating multiple perforated plates with the hole matrix offset. The proprietary material GORE-TEX® has been suggested as a barrier to water but not air, as has sintered polyester weave.

Following the wet CIP process, some installations specify a drying cycle using warm air, to prepare quickly for the SIP phase, which requires that all surfaces be completely free of liquid. This warm air may indeed be delivered through the CIP pipework, thus drying these and the spray heads at the same time. If hydrogen peroxide gassing is to follow, then the ideal scheme is to supply warm, low-humidity air from the isolator air-handling system, thus accelerating the preconditioning phase of the gassing cycle.

Decontamination in toxic and pathogenic applications

Chemical decontamination: cytotoxics

The interior surfaces of isolators used for the processing of known cytotoxic products will become more or less contaminated with the product over time. If cross-contamination of products is to be avoided, and certainly if the isolator is to be opened up for maintenance, a decontamination cycle will be required. In simple isolators, used for minor processes such as dispensing, manual cleaning may be sufficient. In larger and more complex isolators, decontamination systems may be fitted, these being similar to CIP equipment, but placing emphasis on the removal of a particular type of material. This leads to the term *decontaminate-in-place* (DIP). Whether manual or automated, the decontamination process should, if possible, use an agent that deactivates the hazardous product in some way. With some cytotoxic products, a dilute alkaline solution, such as 10 percent sodium carbonate, is suitable. This would be followed by rinsing with water for injection (WFI) to remove residues. The solution drained from the isolator is relatively non-toxic, but is generally treated as still hazardous and should be disposed of accordingly.

Chemical decontamination: unknown hazards

Some research on novel pharmaceutical compounds is carried out in isolators because the potency and toxicity of such compounds are, by definition, unknown. These isolators will need decontamination from time to time; methods similar to cytotoxic isolators apply. If the chemistry of the novel compound is very much an unknown, then perhaps a series of different deactivating solutions might be applied. Ideally, the isolator surfaces should be checked for traces of contamination before opening to the atmosphere.

Biological decontamination: pathogens and recombinant DNA

The containment of organisms in the higher ACDP categories, such as P4 (Ebola virus, green monkey disease, Lassa fever, etc.), is a specialist subject and is beyond the scope of this book. The bovine spongiform encephalopathy (BSE) agent, or prion, presents some extreme and very challenging decontamination problems, also well beyond this text. Standard isolators are regularly used, however, for the containment of lower category pathogens such as tuberculosis, hepatitis, and HIV.

Here, the decontamination emphasis is not only on a cleaning process, but also a sterilisation process. Thus, the various methods for rigorous isolation sterilisation may be applied, such as formaldehyde, PAA, or hydrogen peroxide. These may be used in the liquid or vapour phases, but whichever method is chosen, the method should be validated before routine use. Where waste materials and solutions are removed from biological containment isolators, they should ideally be autoclaved before further disposal.

chapter eight

Running the operation

So far we have considered all of the stages of an isolator project up to checking the work carried out on-site (IQ and OQ), including checking the sterilisation process, where applicable. The next stage is to train the operators and validate the working of the isolator, perhaps first with inert materials and then with the final product process. This is the PQ stage of the project, very much bound up with the term *validation*. The term validation is an all-embracing term and includes IQ and OQ discussed in Chapter 6 and, indeed, really begins with the URS mentioned at the beginning of Chapter 5. Where sterile processing is concerned, validation is mostly taken up with proving that the sterilisation of the isolator is satisfactory. A more general definition of PQ has been given as follows:

Performance Qualification is a two-stage process:

1. Confirmation, with evidence, that the system performs as described in the DQ
2. Confirmation, with evidence, that it continues so to do

In Chapter 5, Figure 5.1 shows the interrelationship of all the stages in the validation process and emphasises the need to address validation early on in any project.

The PQ stage is often a lengthy process in major isolator applications, such as aseptic filling, and usually takes some weeks, often months, and, in some cases, it can take years to complete (Ohms 1996). Curiously, although this stage is lengthy and is vital if a product license is to be obtained, little has been written about the work involved. The book *Isolation Technology* (Wagner and Akers 1995) mentions PQ and validation quite frequently, but really only discusses the issues of sterilisation and then does not go into great depth on the subject. Part of the reason for this may be that validation is specific to one set of equipment used for one specific purpose. Furthermore, it may be specific to one group of operators and even to one regulatory inspector, thus of no great relevance to any other project. Even so, it is possible to make some general remarks about the PQ and validation work.

In aviation circles, it is said that a good landing starts with a good takeoff. Likewise, in pharmaceutical isolation projects, a good validation starts with a good URS, as laid out in Chapter 5. Thus, the final validation of the complete system will be considerably eased if the issues of validation are borne in mind right from the outset. This means getting validation thinking into the design stage. For example, does the ventilation design allow for easy HEPA filter testing? Does the instrument system allow for calibration? If custom software is required to run the isolator through a PC or PLC, then this too will need validation; therefore, it should be written in an appropriate fashion. This point about considering validation from the start is important and should be noted, and actioned, by anyone commencing work on a pharmaceutical isolation project. On a philosophical note, the MHRA (formerly the MCA) and the FDA take the view that validation tests should be run to fail, not run to pass, if they are to challenge the process (Bill 1996). This point is also well covered in a paper by Ken Chapman, which introduces the Proven Acceptable Range (PAR) approach to validation.

Operator training

There is generally a requirement for formal operator training in current GMP (cGMP) regulations, with the following structure:

- Target all the relevant persons, including management.
- Give written, approved programmes.
- Train at the start of the project and continue during operation.
- Set up an assessment protocol.
- Keep accurate, clear records.

Training can start in an informal way at a very early stage, perhaps with an examination of proposal drawings and mockup assemblies where the operators can have a positive input to the design. More formal training best starts at the isolator manufacturer's premises, during FAT. Training could begin with theoretical classroom instruction and be followed up with hands-on work with the actual equipment. Training may continue once the isolator is installed on site and connected to the relevant process equipment and services. Bear in mind that the isolator manufacturer can only give information on its part of the system and that some developmental training will usually take place once the process work has commenced. The isolator manufacturer will normally provide a comprehensive operation and maintenance manual with the equipment, which may form the basis of more formal operator training, if required. The user absolutely must take the opportunity, whilst the manufacturer is on site, to go over all the aspects on the working of the isolator — its construction, control systems and operating parameters, and subsequent maintenance. Indeed, this aspect may be formalised in an agreed package of documentation to be exchanged, possibly even linked to a stage payment for the equipment. Again, this documentation package should be discussed and set out at an early stage in the project, such as the preengineering study.

As part of the training, it may be a good plan to nominate an individual to take charge of the isolator, to become familiar with its workings, and to keep track of its history, maintenance, and repair. At the same time, operator attitude to the work should be monitored and training adapted accordingly. It may well be that the isolator represents a completely different approach to a process that the operators have run for some time by other methods, and so there may be some inertia to overcome. The practice of current GMP (cGMP) should be reemphasised where necessary. Some isolator processes require a sustained high level of mental concentration on the part of the operator, and so an informal and conversational atmosphere should be avoided. The ergonomics of the isolator system will have a big impact on operator attitude, taking us back to the DQ stage once more.

The training stage is also the time when SOPs start to be used — effective training cannot take place without reference to SOPs.

Physical validation

Physical validation is validation of the physical parameters, as opposed to the microbiological parameters, of the isolator system and the resulting isolator environment. It is also termed *commissioning work* in Figure 5.1.

In many ways, physical validation is an extension of the tests carried out as part of OQ described in Chapter 6, and there may be some degree of overlap. Indeed, in many ways, the only difference between OQ and PQ in some tests will be the fact that the process is running, or at least a dummy process is running, for the PQ trials. As with all qualification work, the purpose of the test, the method statement, the results, and the conclusions should be stated and recorded clearly.

Ergonomics and safety

Perhaps the first checks that should be carried out relate to the discussion in Chapter 5 concerning ergonomics. There may be a more formalised series of tests to ensure that all of the operator movements required to perform the isolated process can be carried out without excessive lifting, bending, or turning, as advised by the local health and safety representative. Tests might also be carried out on any safety equipment involved, including such mundane items as raised operator plinths. Any mechanical interlocks on machine access or transfer ports can also be checked at this stage.

CIP functional test

If the isolator has a CIP system fitted, then the next test might be a physical challenge to the CIP system, perhaps using an inert indicator, such as a fluorescing agent mixed with lactose powder. The challenge material would be distributed manually around the entire isolator and then the CIP cycle run. Once the cycle is complete, the interior surfaces of the isolator should be checked for residues of the challenge material. Ideally, a number of cycles would be run using a uniform challenge and varying cycle times. This would establish the cycle time required to reduce residues to a known level; thus, a suitable safety margin could then be added to give a fully validated cycle time. Note that a challenge must be carried out for each individual active product, since they may have widely differing retention qualities.

Particle counting during media filling trials

Particle counting will have been carried out as part of OQ described in Chapter 6, but PQ will normally demand that the same work be carried out during operation, or at least during a simulated process, such as a media filling trial. In addition to the number of particles present, the profile of particles should be mapped, giving the particle size distribution; most particle counters are equipped to do this automatically.

A thorough PQ trial would map a profile of particle sizes at a number of strategic locations in the isolator, such as the filling heads or transfer ports.

Stability of the isolator parameters during production

Stability is perhaps a question of monitoring over time, rather than an active trial under challenge, but the environmental systems should hold conditions within fixed limits over extended periods of time, and data may be needed to demonstrate this situation. Stability might first be tested over a single batch period, perhaps two to three hours of working and then a day-to-day basis for perhaps one to two weeks. The parameters recorded might include isolator pressure, airflow rate, temperature, humidity, and particle count. In this way, the trends of the parameters can be established and incipient failures may be observed before any given parameter actually goes outside its operating limits.

Alarm function

Again, the correct function of any alarm system would be checked during OQ, but it might be further tested as part of PQ since the isolator process may have a bearing on their function.

Mechanical and electronic reliability (GAMP)

A mechanical reliability check is largely a monitoring function, but it should record any failure and describe any remedial action for future reference. Electronic reliability may be bound up with GAMP, as mentioned in Chapter 5, which will address the issue of software validation early on in the project.

To sum up, physical validation should include the following checks:

- Leak test
- HEPA filter integrity
- Air flow rate
- Particle count
- Particle recovery
- Alarm check

Microbiological validation

The cleaning process

Before sterilisation can take place, both during PQ and subsequently during normal production, various cleaning processes will take place apart from any CIP system fitted. These cleaning processes or regimes will need to be validated to show the removal of both nutrient media and active product residues.

The sterilisation process

The checking of the sterilisation process has been mentioned as part of OQ, described in Chapter 6; however, sterilisation is fundamental to any aseptic manufacturing process. A further test might be run as part of PQ, between media fill trials, for instance. This would further reinforce the validation of the sterilisation process, which should ideally be to a known SAL, as described in Chapter 1. The major difference between OQ and PQ here is that OQ would probably be performed on the empty isolator, whilst PQ would be performed with all of the process equipment and containers installed.

Sterilisation of the main isolator is, of course, not the only issue, and the quality of the transfer processes used in operating the system should also be challenged and validated to the same SAL. It is a feature of isolators that they can be validated to a known SAL, unlike cleanrooms, where SAL is highly operator dependent.

Toxic decontamination

In the case of isolators for handling toxic materials, the decontamination process will need to be checked from the safety aspect. Again, this has been discussed as part of OQ, but a continuing trial during the PQ phase would help to establish the validity of the chosen process.

Media fill trials

If the isolator process is an aseptic one, and especially if the process is the filling of product into containers, the real microbiological challenge will be the broth or media fill trial. Andrew Bill of the MHRA (1996) makes the point that, if the media fill produces a contamination rate of 1 in 1,000, then the actual product will also have an infection rate of 1 in 1,000. This means that a major manufacturer producing 1 million units per month will sell 12,000 infected units in a year.

The media-fill batch size and the number of batch runs would be judged in relation to the eventual process, and new guidelines have been issued by the MHRA (Bill 1996) for this type of work. As an example, Novo Nordisk in Denmark (Ohms 1996) ran eight media fill batches, each of 60,000 units, in a new insulin cartridge filling isolator. Since no growth showed in any of these trials, Novo proceeded to full production with a high degree of confidence in the SAL of the filling process (the actual SAL here is better than 4.8×10^{-5}).

It should be noted that the media fill trials are a critical test of the entire project. It is where all the various individual aspects, from equipment design through to operator training and SOPs, are brought together and reviewed closely. A successful media-fill trial; that is, one that shows no growth in the filled containers, is a good indication that all of the factors are correct and that manufacture can proceed.

Viable particle counting during production

Many users will want to check the viable particle count as well as overall particle count, using one of several air sampling methods, such as the slit-to-agar sampler. Ideally, such devices should sample the air without either physical or biological losses and, where unidirectional downflow is operating in the isolator, an isokinetic sampler is preferred. This simply takes the sample at the same velocity as the downflow air. It neither draws excessive air into the sampler nor loses spilled air over the lip of the sampler. The Sartorius MD 8 (Figure 8.1) microbiological air sampler has been described by Parks et al. (1996). This device, using water-soluble gelatin filters with very high particle retention, is ideal for isolator use, and, indeed, some commercial isolators now carry the unit as a standard option. Since the filter can retain very small particles, it has also been used to establish low levels of cytotoxic products in some environments. Figure 8.2 shows the sampler head in use inside an isolator.

Since the isolator is a miniature cleanroom, classic cleanroom techniques of microbiological sampling can be used, such as the following:

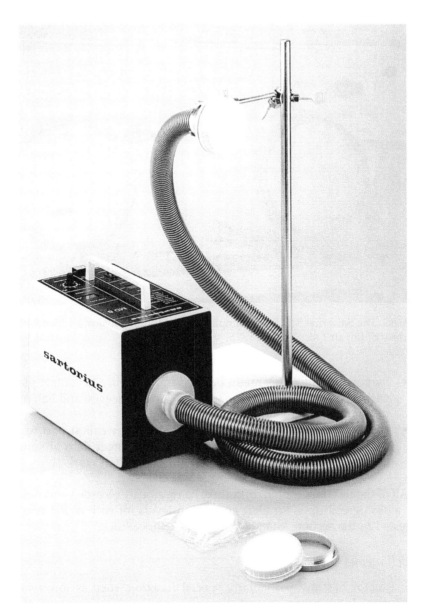

Figure 8.1 The Sartorius MD 8 Air Sampler (Courtesy of Sartorius GmbH.)

- *Settle plates.* These may be laid out at several sites within a large isolator system for specified periods, or perhaps even for the entire period of a production batch.
- *RODAC™ surface testing plates.* These could be applied at critical sites, such as the filling area, but contact plates inevitably leave a residue of media on the surface and this must then be cleaned away carefully after each test.

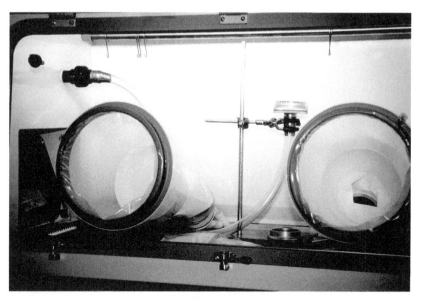

Figure 8.2 The Sartorius MD 8 Air Sampler Head Mounted inside a Glove Isolator (Courtesy of Envair Ltd.)

- *Surface swabbing*. This might be used in preference to contact plates, particularly on the flexible material of gloves, sleeves, and half-suits, to ensure aseptic condition.
- *Finger dabs*. Since the gloves are probably the weakest part of any isolator system, it makes good sense to run finger dabs at suitable intervals, perhaps both before and after a batch of work.

All of the test data should, of course, be rigorously recorded and numerical values agreed upon for alert and action levels for each of the tests, as suggested in the section "Standard Operating Procedures."

Miscellaneous checks

If the isolator has any particularly special function, such as low relative humidity or low oxygen content, then the continued maintenance of the required condition should be tested as part of PQ.

Standard operating procedures

As with any rigorous pharmaceutical process, isolator operations should be run under SOPs, which should be in draft form, at least at the OQ stage, when they can be road-tested ready for approval at the PQ stage. SOPs will set out the precise order of working for each phase of the isolator operation, such as cleaning, sterilisation, transfer, decontamination, and the like. It will describe the tools and materials used and the way in which they are handled.

Any changes to SOPs that become necessary after validation has been completed may need revalidation to ensure continued security. SOPs will probably describe not only routine functions, but also the actions to be taken in the event of any failure taking place. Such events would include immediate emergencies, like a low-pressure alarm trip or the loss of a glove, as well as less urgent situations — for example, the failure of a leak test during servicing. Where possible, numerical values should be assigned to both alert and action levels for any instrument reading or environmental check.

Planned maintenance and servicing

As mentioned in Chapter 6, the manufacturer of the isolator must provide a manual that describes the maintenance tasks that should be carried out to keep the isolator functioning as intended. Some of the routine checks will probably become part of the SOP system.

It may be necessary to plan production around the planned maintenance of the isolator system, possibly coordinating plant shutdown with major maintenance events, such as HEPA filter change.

In some cases, the isolator manufacturer will offer maintenance, either on contract or on demand. In other cases, the operators may be able to provide much of the maintenance themselves, perhaps with the guidance and approval of their QA department. Whichever route is taken, it is a requirement of cGMP to establish how servicing is to be carried out at an early stage in the project, long before a breakdown takes place. Note that media-fill trials may need to be repeated following any significant maintenance, this being part of the change control procedure as mentioned in Figure 5.1.

Documentation

Whatever the process being carried out in the isolator, documentation must be kept up to date. In the case of a licensed product (MRHA or FDA), the documentation actually forms a part of the license, and thus a vital part of production, which must be maintained in a condition ready for examination by the authorities at any time.

As far as the isolator is concerned, this bank of documentation might consist of the following:

- Copy of the URS
- Copy of DQ
- Manufacturer's quotation, client order
- Certificates of conformity, calibration certificates
- FAT

- SAT
- IQ
- OQ
- PQ
- Media-fill trial results
- The operating manual, with drawings and diagrams
- The maintenance manual and maintenance records or logbook; note that the maintenance records are a formal part of any change control system, which must be in place for the entire plant
- The SOPs for the operation of the isolator
- Records of calibration
- Records of operator training
- Records of cleaning
- Change control system description and records
- Isolator logbook

Conclusion

By now, the reader will be aware that the validation process is intimately bound up with the complete project and must be considered from the outset, rather than as a phase that takes place after the equipment has been installed on site. This has been put succinctly, if rather colloquially, in the old cGMP adage: "If it ain't documented, it ain't done."

chapter nine

Regulatory affairs

The MHRA (formerly the MCA)

In common with other regulatory organisations, the MHRA does not issue guidelines or rules as such and does not approve any particular type or make of equipment. It has a remit to examine the production facilities that are submitted for licensing and to report on the quality of the complete process, including all the aspects of validation discussed in Chapter 8.

Thus, the MHRA does not have a view on isolation technology as such. However, individual inspectors have experience that gives them a certain viewpoint. The following account is based on a paper presented at a conference by an MHRA inspector, and reflects his own particular experience and view rather than the opinion or stance of the MHRA.

The MHRA is a British governmental organisation with a remit to monitor and control the supply of all pharmaceutical products to be used in the UK. It has wide-ranging powers, including the issue of both product and manufacturing licenses following detailed inspection of the manufacturer. A large part of their work is concerned with the safe production of sterile materials; thus, the MHRA is encountering isolation technology to an increasing degree. One MHRA inspector's view of isolation technology was presented by Andrew Bill (1996) at the Third Conference of the UK Isolator User Group in Leeds.

The principal area of isolation technology relevant to the MHRA is aseptic production, and perhaps the first point of note is that the MHRA neither encourages nor discourages the use of isolators for this kind of processing. Whilst they acknowledge the potential improvements offered by isolation, they feel that the lack of "an infrastructure of experienced people" means that the application of isolation technology can lead to errors and does not necessarily take advantage of the possible improvements available. In a period of less than one year, the inspector reported witnessing a large number of undesirable situations. The larger, blue-chip companies tended to be less at fault, since they could afford to employ or to train experts, but smaller companies simply lacked the required knowledge and experience.

It has to be said that this is a fairly damning indictment of those who design and manufacture isolation systems and especially of those who operate them. If this really is the experience of the MHRA, then much work remains to be done, even now, to develop an understanding of isolation technology and its practical application.

The MHRA sees the media-fill trial as one of the most significant measures (combined with the other methods of monitoring described in Chapter 8) of the success in excluding microorganisms from a process, and the potential user should note this view and be prepared to provide the required data at the time of inspection. Another suggestion is that validation work should be run to challenge the system properly rather than just produce pass results, thus setting a real challenge to the system and the operators. Again, a clear demonstration of this philosophy to the inspector is likely to gain favour.

The practical considerations

The first practical consideration listed by the MHRA is that of leakage and the establishment of microorganism-proof barriers. The chosen leak test method should give an indication of the size of leak that can be detected, and thus relate to the risk of penetration by microorganisms. Gas testing using Freon and ammonia is thought to have limits of detection around 150 mm, while pressure-hold types of tests may detect smaller holes if the conditions are rigorous enough.

The next consideration is how to exclude challenging microorganisms, raising the question of isolator pressure, as discussed in Chapter 5. Where hazardous materials are in use (cytotoxics and radiopharmaceuticals), safety considerations suggest negative isolator pressure, which in turn will require a high standard for the isolator room if the risk of microorganism penetration is to be reduced. Even so, the use of negative-pressure isolators for aseptic work is to be avoided.

Does positive pressure exclude microorganisms? The MHRA mentions the potential hazards of induction leakage, discussed in Chapter 6 of this book.

Filtration of the airflow through the isolator is the next consideration, and, since sterile filtered air supply is critical to aseptic operation, the MHRA prefers double-inlet HEPA filters for security. Exhaust air must also be HEPA filtered to prevent the return flow of microorganisms; ideally, all HEPA filters should be regularly tested by specialist companies or, at least, by well-trained staff.

The airflow regime is mentioned later in the MHRA inspector's paper under "Considerations about the Process," but it is worth noting here that laminar airflow is considered an enhancement, not a basic requirement of aseptic processing, provided that a true isolator is used and not a less rigorous barrier device.

The process of sterilisation (see Chapter 7) is the next consideration and, in particular, the exposure of all surfaces to the agent in use. Where machinery is used in the isolator, more intractable problems of surface occlusion are created and must be addressed.

The ideal sterilising agent is sporicidal, gaseous, or volatile at ambient temperature and breaks down to innocuous components — hydrogen peroxide and PAA are the main contenders here. The sterilising agent should be introduced upstream of the inlet HEPA filters and removed after the exhaust HEPA filters. Liquid sporicidal agents may be applied to problem areas, such as hinges and dead ends.

The sterilisation cycle must be validated and the process may be lengthy, with the following points being addressed:

- Visualisation of the gas flow.
- Does stratification occur because of density and temperature effects?
- Is condensation a problem?
- Has the agent reached the required concentration in all parts?
- What aeration cycle is required?
- How does the BI used relate to the actual bioburden?
- Does the agent carry over to the BI and give an optimistic result?
- Are the BIs in the most challenging positions?
- What is the effect of isolator loading?
- What safety margin is applied to any critical parameter for the cycle?
- Does the agent penetrate wrappings, stoppers, or containers? What are the implications of this?
- What are the acceptance criteria? Log 6 reduction of a resistant biological indicator, plus a reasonable additional safety margin, is a conservative approach.

Next, the process itself is considered with mention of the following points:

- Aseptic technique should still be used throughout.
- Aseptic connections should be minimised.
- Exposed critical surfaces should be minimised.

Training is seen by the MHRA as critical to the process and must be developed and checked, well structured, and should be extended to all people involved, including the manager.

Ergonomics is seen as important, with comfort, visibility, and lighting all contributing to safe working conditions. Actions in the event of a mishap should be well rehearsed — another point that should be addressed clearly at the time of inspection by the MHRA.

Transfer is a major consideration and needs careful thought. The RTP is seen as more secure than many methods, but not perfect, and still needs treatment, such as manual decontamination of the "ring of concern" (see

Chapter 3) to reduce risk. Large-scale operations that incorporate heat sterilising tunnels or autoclaves need careful management. The continuous output of materials from sterile isolators presents particular problems; generally, batch processes are preferred.

Finally, putting all of these considerations into one philosophy, the MHRA has come up with the acronym PHIGOO for their ideal system — positive, high-integrity, gassed, operator-optimised. The weak points are seen as follows:

- Difficulty in achieving a leak-free boundary
- The limited sensitivity of leak testing
- Difficult cleaning access
- Double door (RTP) transfer systems

Even so, the final statement in the MHRA inspector's paper declares that a PHIGOO system is a quantum leap from many cleanrooms and less rigorous isolator systems with regard to environmental microbiological control.

Table 9.2 from the MHRA shows the infection rates of media-fill trials that result from the various methods of aseptic handling. Clearly, isolation technology has potential benefits that cannot be ignored. The reader will no doubt recognise that many of the points of issue raised in the MHRA inspector's paper by Andrew Bill, on behalf of the MHRA, are addressed in this book.

The U.S. FDA view

The U.S. FDA perspective on isolation technology has been summarised in a brief paper presented by Dr. Ken Muhvich of the FDA Center for Drug Evaluation and Research (CDER) at a conference on isolation held in Rockville, Maryland, in December 1995. Rather than attempt to summarise its contents, it is quoted here in full.

CDER Perspective on Isolator Technology

The FDA recognizes that the aspects manufacture of pharmaceutical products will probably be improved by using isolator filling lines. The frequency of contamination for products manufactured on barrier isolator filling lines should be much lower than those manufactured on conventional filling lines. The rationale is obvious; isolators should prohibit microbial contamination of drug products from personnel and other exogenous sources. However, isolating a filling line raises new concerns regarding sources/mechanisms of contamination for sterile drug products. Hopefully, my comments and the discussion to follow will clarify areas of concern, as well as suggest some practical solutions.

1. Classification of filling room surrounding the locally controlled environment.

We realise that a great portion of the cost savings achieved by installing an isolated line is realised by not having to maintain a cleanroom, i.e., Class 10,000 environment. Several manufacturers and regulators have accepted the use of isolator filling lines in "unclassified, but controlled" areas. By this they mean a room with an air cleanliness class greater than 10,000. By this we hope that they also mean that frequent environmental monitoring of the unclassified area will be performed.

It appears that most manufacturers want to operate the barrier isolator filling lines in Class 100,000 or unclassified areas. We maintain that it might be wise to use the isolator line in a Class 100,000 area, at least initially. The cost savings would still be significant. But what one should use really comes down to what one can validate of an isolator filling line is probably independent of the air cleanliness class of the surrounding area. That is to say, that validation of such a line in a Class 100,000 area is possible. However, one needs to demonstrate that the HEPA filters on the isolator can withstand this type of challenge over time. How long can HEPA filters operate effectively on an isolator in a Class 100,000 or unclassified area? What will the preventive maintenance be on these? How will microbial grow-through at valves, gaskets, rapid transfer ports be prevented? How will gloves be tested for pinhole leaks?

2. Maintenance of sterility inside the barrier — points (weak spots) to consider:

Sterilising filters. There is little doubt that contamination of open product vials from filling room personnel will be negated by using locally controlled environments. However, other potential sources of microbial contamination may become more obvious, e.g., leaks in gloves or HEPA filters. The weak link of an isolator filling line now appears to be the "sterilising filter." Recently, several instances have come to the Agency's attention, when passage of a drug product through a 0.2 micron rated filter was not sufficient to remove contaminating microorganisms. Several issues should be considered: How long will the filling line be operated? Does the subject drug product support the growth of any microorganisms? Does the drug product alter the size of microorganisms or other physical traits, such as deformability, that could affect filterability? How often will the filters be sterilised? A 0.2 micron filter in series with a 0.1 micron filter may be desirable to ensure sterility of the drug solution, especially when any contaminants are exposed to the drug product over a long period of time. A validation certificate from a filter manufacturer alone is probably not sufficient to answer these questions.

Barrier Interface Requirements. Continuous transfer of packaging components from outside the isolator to inside and of product-filled containers from inside the isolator to outside must take place without particle ingress through the mousehole. Maintenance of pressure differentials across the isolator interface is key and a defined pressure cascade should be maintained.

3. Validation of the sterilising method for the area/equipment inside the isolator.

One needs to demonstrate the BIs can be killed in all of the nooks and crannies.

4. Will barrier isolator technology substitute for terminal moist heat sterilisation?

The jury is still out and only time will tell. Until the efficacy of barrier isolator technology is proven to be equivalent, drug products that can withstand the rigors of high thermal input should be terminally sterilized using moist heat.

Sterility Assurance Levels — Many manufacturers of locally controlled filling lines claim a significant increase or "boost" in the Sterility Assurance Level, as compared to the conventional filling lines. In many cases, this is unproven, so they shouldn't claim it. Beyond demonstrating that they can consistently achieve contamination rates of zero percent for media fills of 10,000 vials, nothing has been proven. The first FDA/Barrier Users Group Symposium (BUGS) meeting was held in September 1993. Since that time no one has come forth with a model to relate contamination rates for aseptically processed drug products to sterility assurance levels for terminally sterilized products. The processes are really very different (red apples and oranges) and probably shouldn't be compared in this manner. Also, use of new terminology, such as Sterility Confidence Level (SCL), is probably not helpful, because it has no real basis in science and cannot be used to compare sterilization processes.

It is worth noting that the FDA has a slightly more positive attitude to isolation when compared with the very neutral stance of the MHRA, but they still warn of the possible pitfalls of the new technology. The environmental quality of the isolator room is a very reasonable concern, and it is interesting to note the shift in emphasis given to the product line sterilising filter. Transfer of packaging components in continuous process isolators is another area of concern, as is validation of the sterilising process. These points are very much in parallel with the concerns of the MHRA and so should be addressed carefully by the designers of isolation projects.

Further regulatory comment

James Lyda (1995) has discussed regulatory aspects and makes the very valid point that changes are occurring rapidly in this expanding area of technology. Readers should be alert to any significant changes that may take place. As examples, Lyda cites recent changes to the EU GMP directive, Sterile Products Annex, and the ISO/TC 198 draft on aseptic processing of healthcare products, both of which contain comments on isolation systems.

The relevant regulatory bodies are the FDA CDER in the U.S. and the MHRA in the UK. The European Medicines Evaluation Agency (EMEA) may have influence in the future, and the Pharmacentical Inspection Cooperation Scheme (PICS) has discussed isolation systems; however, at the time of writing, the Japanese Ministry of Health and Welfare (Koseisho) has given no significant guidance.

Lyda then mentions two aseptic production lines that have been approved: at API (Pierre Fabre Médicament Production) in Pau, France, and at Evans Medical in Liverpool, UK. He then goes on to discuss the various aspects of isolation that seem to be the greatest general concerns of the regulators:

- The isolator room environment
- Monitoring of the isolator environment
- Aseptic processing versus terminal sterilisation
- Media fills: value and interpretation
- Weak links in isolation systems

Finally, there is mention of the support that the FDA offers to international standards, provided that they meet certain criteria. The ISO standards meet these criteria; it would seem likely that the documents resulting from the work of ISO/TC 209 and ISO/TC 198 might become the accepted standards of the FDA in the future. This enlightened view should make life easier for those designing and validating isolator projects in years to come.

Standards and guidelines for isolators

The following is a listing of some of the standards and guidelines that apply to isolators. Perhaps the most relevant standard is BS EN ISO 14644, which supersedes a number of former standards, such as the well-known Federal Standard 209 (U.S.) and BS 5295 (UK). At the time of writing, not all of the eight sections are published; indeed, the later sections are only at an early draft stage. However, those published must, of course, be adhered to. Section 7, Separative Devices, is most relevant to those concerned with isolation technology.

The most relevant, or at least the most recent, set of guidelines is that written by the UK Pharmaceutical Isolator Group and due for publication in 2004. This is a comprehensive work based on the practical experience of a wide range of isolator users and covers all aspects from design through to validation, including siting, clothing, physical and microbiological monitoring, leak testing, and the like (see Table 9.1).

Table 9.1 Standards and Guidelines for Isolators

Standard or Guideline Number	Title	Status
BS EN ISO 14644–1 (Cleanrooms and associated controlled environments)	Part 1: Classification of air cleanliness	Published
BS EN ISO 14644–2	Part 2: Specifications for testing and monitoring to prove continued compliance with ISO 14644–1	Published
BS EN ISO 14644–3	Part 3: Metrology and test methods	DIS
BS EN ISO 14644–4	Part 4: Design, construction, and start-up	Published
BS EN ISO 14644–5	Part 5: Operations	FDIS
BS EN ISO 14644–6	Part 6: Terms and definitions	CD
BS EN ISO 14644–7	Part 7: Separative devices (clean air hoods, gloveboxes, isolators, mini-environments)	FDIS
BS EN ISO 14644–8	Part 8: Molecular contamination	DIS in preparation
BS 5295	Environmental cleanliness in enclosed spaces	Parts of this standard are being withdrawn as the relevant sections of ISO 14644 are published
PD 6609: 2000 (UK)	Environmental cleanliness in enclosed spaces — Guide to test methods	An interim document pending publication of ISO 14644, Part 3
BS EN 1822 Parts 1–5	High efficiency air filters (HEPA and ULPA)	Published
UK Pharmaceutical Isolator Group	Isolators for pharmaceutical applications	Due for publication by Pharmaceutical Press Ltd. 2004
PIC/S	Isolators used for aseptic processing and sterility testing	Published June 2000
PDA Technical Report No. 34	Design and validation of isolators systems for the manufacturing and testing of healthcare products	Current version September/October 2001
FDA Monograph	Sterile drug products produced by aseptic processing	Draft of September 2002
PDA BI Working Group	Recommendations for the production, control and use of biological indicators for sporicidal gassing of surfaces within separative enclosures	First draft, June 2002
HSE/MCA (UK)	Handling cytotoxic drugs in isolators in NHS pharmacies	2003

Table 9.2 A Rough Perspective on Safety Arising from the Process of Microbiological Control of Products Labelled "Sterile" (Bill 1996)

Demonstrated Infection Rates from Media Fills	Aseptic Process	Notes
Positives not yet found	Leading edge operations	High grade gassed manufacture
		PHIGOO systems perhaps
1 in 10^5	Modern manufacturing equipment	High speed lines — BFS perhaps
		Use of isolator-like barriers without gassing
1 in 10^4	Conventional manufacturing	Slower
		More interventions
		Use of curtains
	Some CIVAS	Properly run mini-cleanroom-type isolators perhaps
1 in 10^3	Older manufacturing	Low speed
		Manual
		Incomplete shrouds
		Well-run conventional clean room
	The Rest	
1 in 10^2		Complex manual operations
		Not completely closed system
		Some sterility test facilities
		Ambient/ward activity

chapter ten

Case studies

Several papers have been published that give accounts of particular isolation technology projects, especially in the field of pharmaceutical production; some good examples are found in *Isolation Technology* (Wagner and Akers 1995), previously cited in this book. However, it seems appropriate to end this book with two additional examples of isolation in action, to give the reader a flavour of some real-life situations.

The first example is a laboratory-scale operation in flexible film isolators, the second an industrial-scale project in stainless steel enclosures. These two examples were chosen to illustrate the application of the technology at both ends of the scale and, thus, to provide a comparison between the two. Both accounts were written following interviews with the designers and operators of the system described, and were checked for accuracy by these people. The author is very much indebted to these workers for the candid information that was provided. In the interest of impartiality, commercial equipment suppliers have not been named.

Aventis (previously Rhone-Poulenc Rorer), Holmes Chapel, Cheshire, UK

Background

In 1991, the microbiological QA department at RPR was running its sterility testing work in laminar flow cabinets in a cleanroom. As part of a quality improvement programme to increase confidence in sterility testing, alternative technologies were explored. Various options were considered, and a fair amount of research was carried out by talking to other manufacturers and users. Isolation technology was soon accepted by the microbiology QA team as the best way forward.

The first reason for choosing isolation was the prospect of a very significant reduction in the number of false positives, given the very high environmental standards possible inside an isolator. It was hoped that the false-positive rate — a positive growth result that is purely the artefact of the test process — could be reduced to a very low level, perhaps even to zero. Research into existing sterility testing isolators suggested that this might be the case, making isolation very attractive for this reason alone.

Interestingly, the team soon recognised the potential of isolation technology. They felt that a major change was needed from the existing test methods and that isolation would give a boost to the whole department, a new interest, and a new commitment. Whilst they were breaking new ground at the time, this aspect was seen as a benefit, not a disadvantage.

Cost was inevitably a consideration. However, Aventis reported that after taking into account the initial outlay, the installation and commissioning, and, more importantly, the routine operating expenses, there was still a significant savings over refurbishing and running a conventional sterility testing suite.

The flexibility to handle a variety of products was seen as useful, and the ease of operation without full cleanroom gowning was seen as a major advantage. The relative speed with which the isolators could be installed and commissioned also had attractions. It was thought that no other systems of contamination control could offer all of these benefits.

Design and engineering

Reliance was placed on the isolator manufacturers to advise them on the type of equipment, including the methods of sterilisation that should be used for this process. The manufacturer who was chosen from the three interviewed appeared to have the best understanding of the process to be carried out in the isolator.

The isolator system chosen centres around a single half-suit, flexible film, positive-pressure, turbulent flow isolator. It houses a conventional six-funnel membrane filtration manifold, which is connected to the external vacuum supply via a silicone rubber tube that passes through a compression gland in the base-tray wall. The tube is connected to a large Buchner flask to collect liquid waste and provide a water trap, preventing return airflow to the isolator. A two-glove isolator is connected to the main isolator via a simple door on the right-hand side of the suite; it is also connected to an adjacent room by another simple door, passing through the room wall. A further simple door on the two-glove isolator provides for the loading of materials onto stainless steel grid shelves. To the left-hand side of the suite, there is a 310-mm RTP, and both gassable and autoclavable RTP containers were supplied.

The isolators have electronic control for the canopy overpressure and instrumentation of the pressure and airflow rate, with alarms, high and low, audible and visual. The chosen gas generator is an open-loop machine requiring an exhaust duct to atmosphere, which uses the proprietary sterilising solution Citanox as the agent. It consists of 10 percent hydrogen peroxide and 1 percent PAA in the isolator, which achieves a concentration of about

350 ppm by weight. The gas generator has a chart recorder that gives hard copy printouts of the operating parameters for each gassing cycle, which is important for validation purposes.

The original logic was that the main isolator would remain sterile, and the two-glove isolator would be used as a sterilising chamber for the transfer of materials for each batch of work, either from the laboratory or from the adjacent room. The RTP would provide for emergency transfers in the event of an urgent sample. The current practice is to load all of the materials for a batch of work into the main isolator, run a gassing cycle, and then complete the work, before opening the isolator to remove the test materials and repeat the process. The two-glove isolator is also used as a separate unit for any other concurrent work.

Validation

The validation exercise took approximately three weeks to complete. Factory acceptance tests as well as the IQ and OQ activities were carried out by the manufacturer. Essentially, these consisted of HEPA filter tests, leak testing by pressure decay, and instrument calibration. The PQ was written and carried out by the users.

The gassing process was checked using the standard cleanroom techniques of RODAC plates, surface swabs, settle plates, and finger dabs; indeed, these are the environmental monitors used currently on a daily basis. Particle counting was not done, but this did not seem to create a problem with the results or the validation. Total immersion of the ampoules in the medium was also used, and all of the trials were duplicated outside the isolator in laminar flow cabinets to ensure that the same growth resulted.

Both positive and negative controls were run:

- Positive controls: checking that the gassing process did not inhibit the growth of samples known to contain 10 to 100 viable organisms.
- Negative controls: checking that media plates still supported growth after the gassing process.

An operating point to note is that the users did not test the gloves as such, but the gloves were changed after every session. The system has been successfully subjected to both FDA and MHRA inspections.

Conclusions

When asked if the users felt that the decision to launch into isolation technology was a good one, the company gave a firm "yes." The sterility testing results have been very pleasing, and the company has benefited as a result. The company does, however, make the following caveats:

- Isolators do not absolve the user from observing cGMP at all times.
- Documentation is vital. Include service history logs, the entry and exit times of all personnel using the isolators, and log materials entering and exiting the unit. This means that there is full traceability at all times.
- Train the team and encourage them to be proactive in quality improvement and best practice.

The following are direct quotations from the presentation that Aventis now gives quite regularly on their system:

> "Sterility testing is only as good as the people running the system."
> "Training — the people must work as well as the system."
> "Training should be given to all operators, and people's competence and ability should be assessed and recorded."
> "Isolator technology is a unique, extremely rewarding and satisfying way in which an operator can confidently and accurately carry out sterility testing."

Aventis has now installed a second isolator in the same facility on the basis of the excellent results obtained in the previous six years of use. Figures 10.1 to 10.5 illustrate the isolator system. Figure 10.1 is an overall view of the complete isolator system. Figure 10.2 is a more detailed view of the glove isolator, with connection to the main half-suit isolator and through the wall to another room. Figure 10.3 shows the half-suit in the unoccupied condition, suspended by the cuffs. In Figure 10.4, the Millipore Steritest unit is located in front of the half-suit and rests on stainless steel grid shelves to allow the free circulation of sterilising gas. In Figure 10.5, the connection to a further room from the glove isolator is shown.

Dabur Oncology, Bordon, Hampshire, UK

Introduction

In the late 1990s, the large Indian pharmaceutical company, Dabur, recognised a potential for marketing cytotoxic products in Europe. They elected to establish a manufacturing base in the UK and took possession of a former electronic production facility with associated offices, sited in the southeast of England.

The remit for this new facility was the formulation, filling, packing, and shipment of about a dozen different cytotoxic compounds. These were to be available in a variety of different presentations, but with the emphasis on vials in the first stages of the project. There was a requirement for freeze-drying in some cases and for powder filling in others. In short, the facility needed to be very flexible in its operation. At the same time, the facility would require FDA and MHRA licensing and so would need total GMP compliance and full validation throughout.

Figure 10.1 An Overall View of the Aventis Sterility Testing Isolator Suite. (Courtesy of Aventis.)

Figure 10.2 Detailed View of the Glove Isolator. A more detailed view of the glove isolator with connection to the main half-suit isolator and also through the wall to another room. (Courtesy of Aventis.)

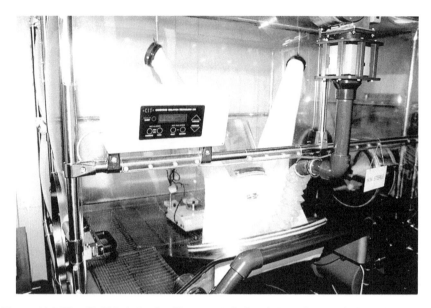

Figure 10.3 The Half-Suit in the Unoccupied Condition, Suspended by the Cuffs. (Courtesy of Aventis.)

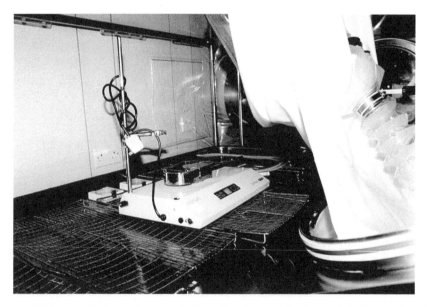

Figure 10.4 The Millipore Steritest. This unit is located in front of the half-suit and rests on stainless steel grid shelves to allow free circulation of the sterilising gas. (Courtesy of Aventis.)

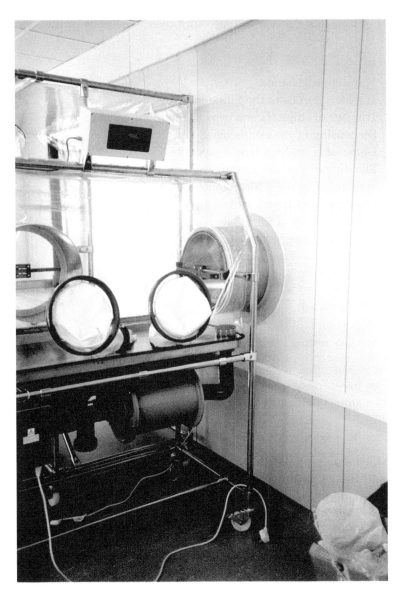

Figure 10.5 The Connection of the Glove Isolator to a Further Room is Shown in More Detail. (Courtesy of Aventis.)

Beyond the GMP aspects of the work, Dabur (pronounced to rhyme with *harbour*) were very conscious of the safety implications for operators handling relatively large quantities of cytotoxic actives on a daily basis. They were also looking for an advanced plant with high prestige value. These factors of GMP, safety, and appearance combined to make the use of isolators to contain the process work an obvious solution.

The production process for each formulation consists of the following stages:

- Vial washing
- Vial depyrogenation
- Transfer of vials to the filling machine
- Transfer of stoppers and caps to the filling machine
- Dispensing of the excipients
- Dispensing of the active compounds
- Blending of powder formulations
- Transfer of powder formulations to the filling machine
- Dissolution of liquid formulations
- Transfer of liquid formulations to the filling machine
- Vial filling, stoppering, and capping (if not freeze-dried)
- Vial filling and partial stoppering (if freeze-dried)
- Freeze-drying
- Capping
- Exterior wash
- Inspection
- Sterility testing

It was clear that isolation would be required to contain all of these processes except inspection. It was also clear that gas-phase sterilisation would be almost obligatory if MHRA approval were to be gained.

Positive- or negative-pressure isolators?

One of the first questions to be tackled was that of isolator operating pressure — should this be positive for product protection or negative for operator protection? After consultation with the MHRA (then the MCA) and the HSE, the conclusion was that the isolators should be positive-pressure where aseptic manufacturing conditions were required. It was deemed that the risk to operators from the bulk liquid was relatively low — it is powder material that presents a greater hazard. If the isolators were built and maintained to a high standard of containment, then positive pressure presented little risk to the operators.

The only exceptions were to be the dispensing and formulation isolators, because these deal with the actives in the powder form and are not aseptic operations. This, then, was to be under negative pressure. After some two years of operation, this choice of pressure regime has proved to work satisfactorily.

General layout and cleanrooms

Given the use of positive-pressure isolators, Dabur could have opted, with some justification, for grade D cleanrooms. However, in the absence of any guidelines or consensus at that time, and, for the relatively small extra cost involved, they elected to use grade C throughout the production area.

The basic layout of the facility evolved quite quickly. The first layouts showed an inspection corridor around the entire production area, but space constraints soon reduced this to a corridor around only two sides (see Figure 10.6). A .large room provides for the main filling process with access or

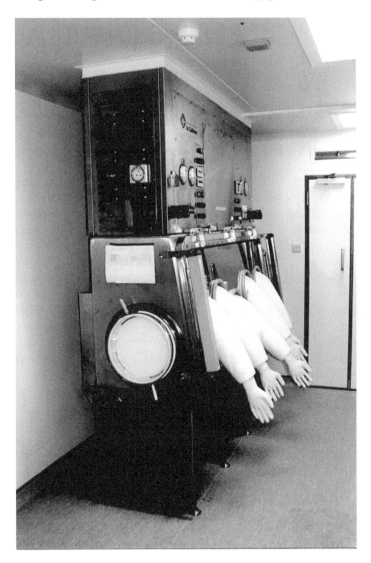

Figure 10.6 General View of the Dispensing Isolator Showing the 350-mm RTP with a Dummy Container Docked in Place Ready for Gassing. The door of the RTP will be opened during gassing and thus the occluded surfaces of the port will be sanitised. Note also the vertical bars of the light beam array interlock on either side of the front window.

interface through to the depryogenation oven, autoclave, freeze dryer, and vial exterior washing machine. There are separate rooms for dispensing, formulation, vial and vessel preparation, and inspection.

The rooms run at positive pressure, with a cascade through from the highest pressure in the filling area (35 Pa) to the lowest pressure in the inspection area (10 Pa). Once again, it was felt that the risk to the outside world from the loss of toxic material already contained within isolators was minimal. The only exception here is the dispensary, which runs at a slight negative pressure in order to avoid possible spread of the toxic powders that are handled in this area

Filling line design

The initial designs for the filling line called for a fixed system of isolators in a line, through from the depyro oven to the vial exterior washer. This then required a branch off to the freeze dryer, with a return line either to the filling machine for capping or to a separate capping machine. Furthermore, there was a requirement to cap powder-filled vials produced in the dispensing isolator. When the future requirement for a syringe filling line was added, the line became complex. A further problem arose with the access to both sides of a line that stretched from one end of the cleanroom to another. Neither a bridge nor a tunnel was practical, nor was leaving the cleanroom and then reentering on the other side.

Thus, a system of mobile transfer isolators was developed. It was quickly realised that a mobile isolator capable of carrying enough vials for the largest specified batch would be heavy — indeed, it would weigh around half a tonne. Could such an isolator be moved around the complex structures of the cleanrooms? The isolator manufacturers produced a full-scale mockup in MDF, running on a total of ten cleanroom castors. They loaded it with sandbags to the maximum projected weight. It was found that two operators could safely move this isolator with sufficient precision to dock with any of the workstations.

There are therefore three fixed isolators and one port to deliver filled vials through the cleanroom wall to the vial exterior washing and inspection machines. Batches of trays, up to 36 at a time, are moved between the fixed isolators using any of three mobile isolators. La Calhène SA 350-mm RTPs are used to dock the isolators together, giving an immediate sterile transfer. The trays have been specifically produced to pass through the 350-mm RTP when loaded with the largest vial size required.

General isolator design

The fixed isolators are all fabricated from 316L stainless steel and fitted with glass or Perspex windows. Either gloves or half-suits are used for access. Air is taken from the room, passes double inlet HEPA filters as specified by the MHRA, circulates through the isolator, and is returned to the room via a single exhaust HEPA filter.

The airflow regime specified is turbulent at up to greater than 60 air changes per hour. Unidirectional downflow has not been specified, as it is not uncommon in isolator applications, even in the filling isolator. The use of what has been termed *engineered airflow* requires the bespoke design of machines inside the isolator to work effectively, but the application of unidirectional downflow has major cost implications, both for initial plant costs and subsequent running costs. The only significant source of particles in a sealed isolator is the process itself and, with good cleaning and sterilisation technique, these will be nonviable. Given careful design, particle generation can be low and the direction of flow engineered to move the burden away from critical areas. Such logic led to the choice of turbulent airflow in these isolators.

The mobile isolators have stainless steel base trays but have clear flexible film canopies to reduce the overall weight.

The isolators all have active control of the set operating pressure. They have instrumentation and alarm of the airflow rate and the internal pressure. They all have HEPA filter test ports fitted to the front panels to make testing a simple process. They also have gassing connections and valves fitted to accessible panels on the front or end faces.

A full-size mockup of each isolator was made by the manufacturers so that ergonomics could be checked and developed. In the case of the filling isolator, the mockup was of sufficient quality to operate and test the airflow pattern.

Specific isolator designs detail

Dispensing isolator

This features a well to accommodate the balance. It carries a powder-filling machine and it is also fitted with blending capability for powder products. The drive of the blender is fixed outside on the back wall of the isolator and passes through a shaft seal to the interior. A 10-kg drum can be filled with powder and loaded onto the blender within the isolator. The operator is prevented from entering the isolator sleeves during blending by a beam array across the front face. If this beam array is broken, the blender stops immediately.

Bulk material is delivered to the isolator, and dispensed batches are delivered from the isolator via RTP containers.

Formulation isolator

The formulation isolator is fitted with two stainless steel vessels that bolt up under the base tray, one of 200-litre capacity and one of 60 litres. Each vessel has a heating element and a cooling jacket, together with magnetically coupled stirring paddles.

A supply of WFI is directly piped into the isolator, delivered by an automatic batching device. Thus, the operator dials up the required volume of WFI and this is delivered to the formulation vessel directly. The formulation isolator has facility for nitrogen blanketing, since some of the products are formulated in solvents.

Once formulated, the product is pumped through a sterilising filter into a mobile stainless steel vessel that has been previously SIP sterilised. This includes the delivery hose housed within a special stainless steel RTP container. The product vessel can thus be docked onto the filling isolator and connected to the filling machine aseptically, using the RTP.

Depyrogenation isolator

Trays of washed vials are loaded into the depyro oven in the preparation area. The oven carriage delivers these trays into a half-suit isolator, whose operator passes them through to a docked mobile isolator. They can then be transferred to the filling isolator.

Mobile isolators

These are fitted with a rack system that can be raised and lowered by pneumatic rams. Each shelf can be presented to a "railway" that carries the vial trays; thus, the operator does not have to lift the trays but only push them along the railway and onto each shelf in turn.

Filling isolator

The filling isolator is mounted onto the extended bedplate of an Inova vial filling machine. This machine is basically standard, but has been adapted to mate with the isolator. The manufacturers of both items were required to work in close cooperation and, in the event, there were few problems when the two came together on site. Both the isolator and the filling machine were leak tested separately at FAT, a special flexible film canopy being made for the filling machine test.

The tray infeed arm of the filling isolator can accommodate one tray, and thus an operator can keep the vial accumulation table continuously stocked from the mobile isolator. Empty trays are passed manually across to the output arm of the isolators, where they are then loaded with filled vials. Full trays are finally pushed out to another receiving mobile isolator.

Stoppers and caps are filled into La Calhène SA BetaBags in the preparation area. These are 190-mm RTP container flanges fitted with Tyvek bags, so that the complete assembly can be autoclaved on porous load cycle. The bags are then docked onto the filling isolator, and the stoppers and caps move down delivery chutes into the vibratory bowl feeders.

The airflow pattern was tested and developed using the full-size mockup mentioned earlier. Air is fed to the isolator interior by three ceiling-mounted HEPA filters, arranged so that the direction of flow is away from the filling area and toward the infeed and output arms of the isolator, from where the air is exhausted. The air change rate is specified at 60 per hour, and this engineered airflow produces particle counts meeting Class A conditions of the Orange Guide (GMP) in operation. Thus, the complexities and expense of unidirectional downflow have been successfully avoided in this installation, although more recently there has been increased regulatory pressure to apply downflow under these circumstances.

Freeze-dryer isolator

The main problem with both autoclaves and freeze dryers linked to isolators is the door. Sliding doors are difficult to sterilize; hinged doors are large and heavy. In this case, the freeze dryer is not very large, and so the isolator could be engineered so that the door simply opens out into the isolator. An ergonomic study was made using a full-size mockup, and the half-suit has been placed so that the operator can open the dryer door safely and still access the interior to load trays. Again, trays are presented from a mobile isolator and returned to the mobile when the cycle is complete.

The isolator is sealed to the face of the freeze dryer with a flange, in effect a short tunnel, supplied by the isolator manufacturer. Some difficulty was experienced in getting the face of the freeze dryer, which is presented inside the isolator, leak-tight since the equipment design did not fully interpret the requirements. This was eventually overcome by locating and sealing specific leak paths.

Gas generators

At the time of design for this facility (and, indeed, at the time of writing this account) there was a choice of just two gas generators for isolator sterilisation. These were the Steris (formerly AMSCO) VHP 1000 and the BIOQUELL Clarus C. These machines were similar in size and performance, but the BIOQUELL unit appeared to have some technological advantages (see Chapter 7), though it was very new to the market. The Steris unit, on the other hand, had almost ten years of working experience in the industry. In the final analysis, however, the lower cost of the BIOQUELL unit swayed the decision.

Three Clarus units were specified, two for the production area and one for the sterility testing laboratory isolators. Reliability problems were experienced in the early years of using the generators, but these have been largely ironed out as experience has been gained, both by Dabur staff and BIOQUELL engineers. Some doubts still remain concerning the reliability and calibration of the hydrogen peroxide measurement probes.

Sterility testing

Two flexible film isolators were specified for the sterility testing laboratory. One of these acts as a transfer isolator and is gassed for each batch of work. The other carries a Millipore Steritest unit and acts as the working isolator.

This has become the classical arrangement for sterility testing isolators. The system has worked well for Dabur.

Validation

An extensive programme of validation was set up from the start of the project. Protocols were mostly written by the Dabur QA team and executed variously by Dabur technicians and the equipment suppliers. The process has been quite long but has been satisfactorily completed. A change from gas injection through the HEPA filters to direct gas injection into the isolators improved the gassing process significantly and thus eased the validation of the gassing cycles.

Review in the light of experience

Some three years after the facility was designed, does it meet the expectations of the owners? Are they broadly satisfied with the design or would they make radical changes, given the opportunity?

The answer from the Dabur team is a qualified "yes," they are satisfied. The multifunction requirement, combined with isolation technology, made things less easy. Had the processes been known more clearly, then the equipment might have been better arranged and the ergonomics consequently improved. Dabur also found that equipment and machinery suppliers professed to understand the interface with isolators, but in fact lacked experience and expertise.

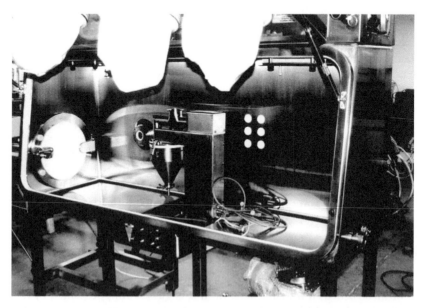

Figure 10.7 Interior of the Dispensing Isolator Showing the Powder Filling Machine in Place. The drive of the blending machine is visible just to the left of the filler

Figure 10.8 The Compounding Isolator. Note the 200-litre and 60-litre vessels mounted under the floor. These are fitted with wheels to aid removal during major cleaning,
but are normally jacked up to seal against the floor of the isolator.

Figure 10.9 The Depyrogenation Oven Isolator. The carriage moves out of the oven to occupy the space immediately in front of the half-suit operator. The loading end of the oven is accessed from the preparation area, which is located through the double doors on the left.

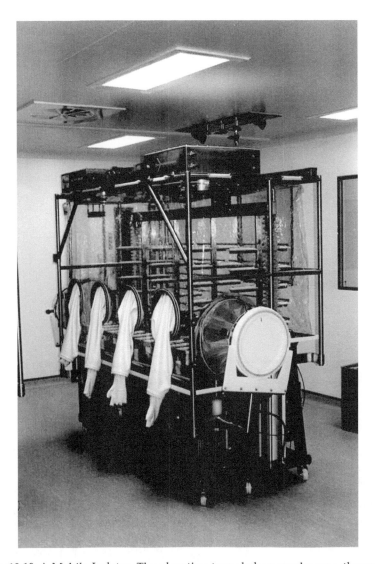

Figure 10.10 A Mobile Isolator. The elevating tray shelves can be seen through the flexible film canopy. The tray railway is seen as the white rollers, just below the shoulder rings, extending to the male (beta) RTPs on each end of the isolator.

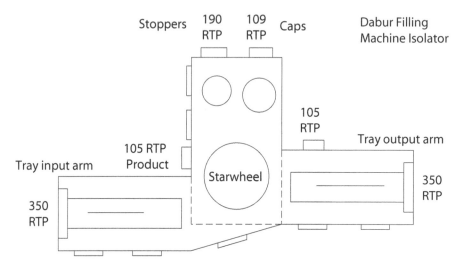

Figure 10.11 Schematic Drawing of the Vial Filling Isolator Layout.

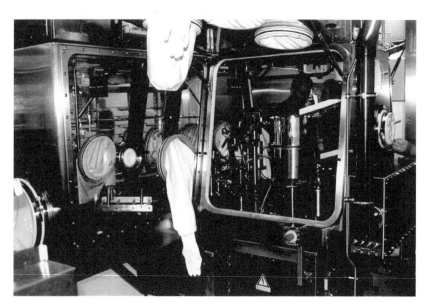

Figure 10.12 A View of the Vial Filling Isolator from the Northeast of the Schematic Drawing.

Figure 10.13 A View into the Vial Filling Isolator from Due South of the Schematic Drawing. Note the two RTPs at the back for introducing stoppers and caps, the inlet HEPA filter on the roof, and the gas distribution fans, also on the roof.

Figure 10.14 An SIP RTP Container as Used for Making Aseptic Connection between the Mobile Product Vessels and the Filling Isolator. The flying saucer-like object fits onto the container during SIP, allowing the lid to be lifted and thus getting steam to the occluded surfaces of the lid assembly.

Figure 10.15 The Freeze-Dryer Isolator. The half-suit has been removed so that the door of the drier can be clearly seen in this view. It hinges into a recess on the back face of the isolator, opposite the half-suit plinth.

References

Advisory Committee on Dangerous Pathogens (ACDP), Guidance on the Use, Testing, and Maintenance of Laboratory and Animal Flexible Film Isolators, published by HMSO for the Health and Safety Executive, 1985.

AECP 1062, The Parjo Method of Leak Rate Testing Low Pressure Containers, published by the Standards Section, Engineering Directorate, UKAEA (Northern Division), 1981.

Agalloco, J.P., Isolation technology: A consultant's perspective, in *Isolator Technology*, Wagner, C.M. and Akers, J.E., Eds., Interpharm Press, Buffalo Grove, IL, 1995, p. 324.

Agalloco, J.P., Microbiological Quality Assurance in Pharmaceutical Isolators, Proceedings of the Scottish Society for Contamination Control Meeting on Pharmaceutical Isolators, Birmingham, UK, 1996.

Akers, J.E., Isolator technology: Regulatory issues and performance expectations, in *Isolator Technology*, Wagner, C.M. and Akers, J.E., Eds., Interpharm Press, Buffalo Grove, IL, 1995, pp. 45–48, 51.

Bill, A., Some Considerations about Aseptic Processing and the Use of Isolators, paper presented to the Third Isolator Conference of the Isolator User Group, Leeds, UK, 1996.

BS 5295:1976, Environmental Cleanliness in Enclosed Spaces, British Standards Institution.

BS 5295:1989, Environmental Cleanliness in Enclosed Spaces, British Standards Institution.

BS 5726:1979, Specification for Microbiological Safety Cabinets, British Standards Institution.

BS EN 1822:2000, High Efficiency Air Filters (HEPA and ULPA).

BS PD 6609:2000, Environmental Cleanliness in Enclosed Spaces — Guide to Test Methods.

Davenport, S.M. and Melgaard, H.L., Ultraviolet Pass-Through as a Transfer Technology in Barrier and Isolator Systems, paper presented at the PDA/ISPE Joint Conference on Advanced Barrier Technology, Atlanta, GA, 1995.

EC Guide to GMP for Medicinal Products, published by HMSO for the MHRA ("European GMP"), 1992.

EN 12469:2000 E: Biotechnology — Performance Criteria for Microbiological Safety Cabinets, Brussels.

Farquharson, G.F., British and European experience with isolator technology, in *Isolator Technology*, Wagner, C.M. and Akers, J.E., Eds., Interpharm Press, Buffalo Grove, IL, 1995, pp. 79, 91–96.

Farwell, J., *J. Hosp. Infection*, 27:263–273, 1994.

Frean, J.A., Probable Causes of the Johannesburg Fatalities, Proceedings of the Scottish Society for Contamination Control Meeting on Pharmaceutical Isolators, Birmingham, UK, 1996.

GAMP, Good Automated Manufacturing Practice: Supplier Guide for Validation of Automated Systems in Pharmaceutical Manufacture, produced by the GAMP Forum for the International Society for Pharmaceutical Engineering, 1996.

Health & Safety Executive, Handling Cytotoxic Drugs in Isolators in NHS Pharmacies, HSE/MHRA, 2003.

ISO/DIS 14644-7: Cleanrooms and Associated Controlled Environments — Part 7: Separative Enclosures (clean air hoods, glove boxes, isolators, and minienvironments), IEST, Geneva, 2000.

Lee, G.M. and Midcalf, B., Eds., *Isolators for Pharmaceutical Applications* (*Yellow Guide*), published by HMSO for the UK Pharmaceutical Isolators Group, 1994.

Lumsden, G., personal communication, 1996.

Lyda, J.C., Regulatory aspects of isolation/barrier technology, in *Isolator Technology*, Wagner, C.M. and Akers, J.E., Eds., Interpharm Press, Buffalo Grove, IL, 1995, p. 67.

MDH Ltd., technical literature, 1995.

Melgaard, H.L., Barrier Isolator Design Considerations for Clean-in-Place, presentation to the PDA/ISPE Joint Conference on Advanced Barrier Technology, Atlanta, GA, 1995.

Meyer, D., Developing a barrier/isolator implementation plan, in *Isolator Technology*, Wagner, C.M. and Akers, J.E., Eds., Interpharm Press, Buffalo Grove, IL, 1995, p. 116.

Michael, D., International Standards — Current Developments, paper presented to the Third Isolator Conference of the Isolator User Group, Leeds, UK, 1996.

Midcalf, B.M., Phillips, W.M., Neiger, J.S., and Coles, T.P., Eds., *Pharmaceutical Isolator Working Party, Pharmaceutical Isolators and their Application* (3rd edition of the *Yellow Guide*), awaiting publication by The Pharmaceutical Press, London, May 2004.

Muhvich, K., Perspective on Isolator Technology, presented at the Barrier Isolation Technology Conference, Rockville, MD, 4–6 December, 1995.

Neiger, J., Life with the UK pharmaceutical isolator guidelines: a manufacturer's viewpoint, *Eur. J. Parenteral Sci.*, 2(1): 13–20, 1997.

Neiger J.S., Negative Isolation, Cleanroom Technology, April 24–25, 2001.

Ohms, R., Assembly and Filling of Cartridges in an Isolator, Proceedings of the R3 Nordic Meeting, Copenhagen, 1996.

Parks, S.R., Bennett, A.M., Speight, S.E., and Benbough, J.E., An assessment of the Sartorius MD8 microbiological air sampler, *J. Appl. Bacteriol.*, 80:529–534, 1996.

PDA (U.S. Parenteral Drug Association) Technical Report 34: Design and Validation of Isolator Systems for the Manufacturing and Testing of Healthcare Products, Bethesda, MD, 2001.

PDA Biological Indicator Working Group (Europe), Recommendations for Biological Indicators Used in Development and Qualification of Sporicidal Gassing Processes, in preparation, 2004.

Pflug, I., Sterilizing the Inside of the Barrier Isolator Using Atmospheric Steam Plus Hydrogen Peroxide, paper presented at the PDA/ISPE Joint Conference on Advanced Barrier Technology, Atlanta, GA, 1995.

PIC/S (Pharmaceutical Inspection Convention — Pharmaceutical Inspection Co-Operation Scheme), Isolators Used for Aseptic Processing and Sterility Testing, PIC/S Secretariat, Geneva, 2002.

Rahe, H.D., Advanced barrier isolation technology saves cost/space, in *Clean Rooms*, Haystead, J.S., Ed., PennWell Publishing, Nashua, NH, 1996, 10:27–30.

Schumb, W.C., Satterfield, C.N., and Wentworth, R.L., Hydrogen Peroxide, published by Reinhold Publishing for the American Chemical Society, 1956.

Sterile Drug Products Produced by Aseptic Processing — Current Good Manufacturing Practice, Food & Drug Administration (FDA), draft guidance , Rockville, MD, 2003.

Thomas, P.H. and Fenton-May, V., *Pharmaceutical J.*, 238:775–777, 1994.

U.S. Federal Standard 209E, 1992.

Wagner, C.M., Current challenges to isolation technology, in *Isolator Technology*, Wagner, C.M. and Akers, J.E., Eds., Interpharm Press, Buffalo Grove, IL, 1995, p. 61.

Wagner, C.M. and Akers, J.E., Eds., *Isolator Technology*, Interpharm Press, Buffalo Grove, IL, 1995.

Watling, D., Ryle, C., Parks, M., and Christopher, M., Theoretical analysis of the condensation of hydrogen peroxide gas and water vapour as used in surface decontamination, *PDA J. Pharm. Sci. Tech.*, 56:8, 2002.

White, P., personal communication, 1998.

Glossary

ACDP Advisory Committee on Dangerous Pathogens. A committee in the UK to advise on the use and handling of pathogenic materials.

AECP Atomic Energy Commission Protocol.

Arimosis The "leak-tightness" of an isolator.

BI Biological Indicator. Typically resistant spores, such as *Bacillus subtilis* or *B. stearothermophilus*, used to test efficacy of a sterilising agent.

Biomedical A euphemism used, in a sensitive political climate, to denote research animal applications.

Breach Velocity The velocity of air passing through an inadvertent breach in the wall of an isolator, designed to prevent the ingress of contamination to an aseptic isolator or the loss of toxic material from a containment isolator. Provided that this velocity exceeds 0.70 m/sec, no loss is deemed to take place. The conventional test breach diameter is 100 mm, representing the loss of a glove.

CIP/SIP Clean-in-Place/Sterilise-in-Place. The automated process of cleaning and then sterilising an isolator or other piece of pharmaceutical equipment.

CIVAS Central Intravenous Additive Service

Contamination Unwanted material in the air, in process materials, or on surfaces. For the purposes of this book, contamination may be microbiological, particulate, or chemical.

DIP Decontaminate-in-Place. Similar to CIP, but referring to toxic rather than biological contamination.

DOP Dioctyl Phthalate. The chemical originally used to form smokes for air filter testing. Now thought to be toxic, it is substituted with food-grade oils and the acronym is translated to *dispersed oil particulates*.

DPTE From the French *double porte de transfer etanche*; the double-door transfer system used on isolators. See also RTP.

DQ Design Qualification. The final specification document issued by the design team of a new pharmaceutical plant project, describing what is to be built and how it is to operate.

False Positive In sterility testing, a false positive is a sample showing the presence of microorganisms that result from the experimental process, not from the sample itself.

FAT Factory Acceptance Tests. Tests carried out on a new piece of equipment such as an isolator, within the manufacturer's premises, before delivery.

FDA Food and Drug Administration. The governmental body that regulates the production of pharmaceuticals in the U.S.

GAMP Good Automated Manufacturing Practice. Similar to GMP, but applied to automatic processes.

Gantt Chart A diagram giving a visual summary of the stages of a project, such as the setting up of a new isolator suite.

GLP Good Laboratory Practice. Similar to GMP, but applied to laboratory-scale operations.

GMP Good Manufacturing Practice. The concept of optimal design and operation applied to pharmaceutical processes, particularly those of an aseptic nature. The most recent issue is referred to as cGMP (current GMP).

HEPA High Efficiency Particle Air (Filter). A type of air filter, usually defined as having greater than 99.997 percent efficiency at 0.30-mm particle size. Often made from glass-fibre paper, it may be regarded as providing sterile air.

HPV Hydrogen Peroxide Vapour. H_2O_2.

HSE Health and Safety Executive. The governmental body that regulates industrial safety in the UK

HVAC Heating, Ventilation and Air Conditioning. The system used to control the quality of air, in terms of temperature and humidity, in an isolator or other air system.

IBC Intermediate Bulk Container.

IMS Industrial Methylated Spirits. A simple surface disinfectant consisting of 70 percent ethanol in water with the addition of methyl alcohol to denature the solution.

IPA Isopropyl Alcohol. A simple surface disinfectant consisting of a 70 percent solution of alcohol.

IQ Installation Qualification. A document that essentially lays out how and where a new piece of equipment, such as an isolator, will be installed.

Isolator Variously defined but, for the purposes of this book, denotes a working area separated from the operators by either a physical wall or a HEPA filter at all times, including during any transfer process. Certain types of engineered airflow transfer devices are, however, acceptable within the remit of this book.

Laminar Flow The former term for *unidirectional flow*.

MCA Medicines Control Agency. The governmental body that regulates the production of pharmaceuticals in the UK, now known as the MHRA.

MHRA Medicines and Healthcare Products Regulatory Agency. The governmental body that regulates the production of pharmaceuticals in the UK. Formerly known as the MCA.

NAMAS National Measurement Accreditation Service. The scheme used in the UK to provide calibration services back to national standards.

OEL Occupational Exposure Limit. The allowable exposure of workers to an active substance, defined usually in terms of the concentration permitted for defined time periods.

OEM Original Equipment Manufacture. The use of standard components from one manufacturer in the products of another.

OQ Operational Qualification. A document that lays out what tests will be applied to a newly installed piece of equipment such as an isolator in terms of its operating characteristics, before active process work begins. The tests typically consist of HEPA filter checks, leak tests, flowrate checks, instrument calibration, and the like.

PAA Peracetic Acid. $CH_3C(O)OOH$.

Parenteral From the Greek para = beside and enteron = the alimentary tract. In effect, this term is used to refer to drugs that are administed by infusion or injection.

PHIGOO Positive-pressure, High-Integrity, Gas-sterilised, Operator-Optimised. The term coined by a member of the MHRA to describe the favoured form of isolators for aseptic operation.

P&ID Piping and Instrumentation Diagram. The drawing showing all of the ductwork concerned with ventilation, CIP/SIP, services, instrumentation, and the like associated with an isolator system.

PQ Performance Qualification. A document that lays out the tests carried out on a piece of equipment such as an isolator in terms of its performance in active process work. The tests typically consist of checking the sterilisation system with biological indicators, particle counting, settle plate counting, and media fill trials.

QA Quality Assurance. The process within a pharmaceutical company, usually handled by a specifically dedicated department, with a remit to regulate product quality.

QC Quality Control. A similar process to Quality Assurance.

Ra Roughness average. A measure of surface finish.

RF Radio-Frequency. Flexible film canopies are fabricated by a welding process that involves the use of a radio-frequency-tuned power circuit.

RTC Rapid Transfer Container. A container to be used with the Rapid Transfer Port system.

RTP Rapid Transfer Port.

SAL Sterility Assurance Level. The probability of finding a nonsterile product within a given batch of product.

SAT Site Acceptance Tests. Tests carried out on a new piece of equipment such as an isolator, within the user's premises after delivery.

SOP Standard Operating Procedures. A document that lists the standard, validated sequence of actions to be used to perform a process.

SPF Specific Pathogen Free. Animals that are free from certain listed pathogens.

TPN Total Parenteral Nutrition. The feeding of patients entirely by the infusion of a suitable blend of basic nutrients.

Transfer Port The blanket term used in this book to define any mechanism for transfer into or out of an isolator.

ULPA Ultra Low Particulate Air (Filter). Similar to the HEPA filter but having greater efficiency, defined as 99.9999 percent efficient at 0.30-micron particle size.

Unidirectional Flow An airflow in which all of the stream lines are parallel and no turbulent mixing exists. Such an airflow will not entrain contamination from outside the region of flow. The conventional specification for unidirectional flow in air is 0.45 ± 0.09 m/sec. See also Laminar Flow.

URB User Requirement Brief.

URS User Requirement Specification. The initial document issued by the design team at the start of a new pharmaceutical plant project, used to describe broadly what is required of the new plant.

VHP Vapour of Hydrogen Peroxide. The trademark of the AMSCO corporation.

WFI Water for Injection. Water of a sufficient purity to be used for injection or infusion.

Index

A

ACDP, *see* Advisory Committee on Dangerous Pathogens
Acquired immune deficiency syndrome (AIDS), 18
Acrylic, 24
Activated carbon filters, 32–34
Advisory Committee on Dangerous Pathogens (ACDP), 124
AIDS, *see* Acquired immune deficiency syndrome
Air
  handling, 26, 43
  purging, 54
  sampling, 94
Airborne bacteria, 29
Air conditioning, 41
Airflow
  patterns, 133
  schematic, 41, 42
Air suits, 9
AISI, *see* American Iron and Steel Institute
Alarm(s), 135
  function, 166
  tests, 116
Ambidextrous gloves, 48
American Iron and Steel Institute (AISI), 24
Anemometers, 133
Arimosis, 23, 123
Asepco Radial Diaphragm Valve™, 110
Aseptic dispensing, 19
Astec Microflow Citomat, 144, 145, 146
Autoclavable containers, 69, 70
Autoclaves, 8, 77
Aventis, 183

B

Bacillus
  stearothermophilus, 138
  subtilis, 137, 138, 149, 158
Bacteria, HEPA filter and airborne, 29
Bagging port, 58
Bag-over-bag techniques, 62
Barrier technology, 2, 4
Barrier Users Group Symposium (BUGS), 178
Battery backup, isolator, 72
Bernoulli's law, calculation using, 132
Beta Bag™ system, 69
Bio-decontamination, 151, 153
Biological decontamination, 161
Biological safety cabinet, biomedical isolators connected to, 84
Biomedical isolation, 16
Biomedical isolator(s)
  connection of to biological safety cabinet, 84
  monitoring system fitted to, 89
  PVC sleeves on, 47
BioQuell Clarus™, 151, 152, 154, 155
Blocking down, 22
BMS, *see* Building management system
Bovine spongiform encephalopathy (BSE), 161
Breach velocity, 37, 116, 135
Brownian motion effect, 27
BSE, *see* Bovine spongiform encephalopathy
Buck port, 75, 76
BUGS, *see* Barrier Users Group Symposium
Building management system (BMS), 94

C

CAD, *see* Computer-aided design
Canopies, flexible film, 22
Case studies, 183–204